Gerardus Beekman Docharty

Elements of Analytical Geometry and of the Differential and Integral

Calculus

Gerardus Beekman Docharty

Elements of Analytical Geometry and of the Differential and Integral Calculus

ISBN/EAN: 9783337812850

Printed in Europe, USA, Canada, Australia, Japan

Cover: Foto ©berggeist007 / pixelio.de

More available books at **www.hansebooks.com**

ELEMENTS

OF

ANALYTICAL GEOMETRY,

AND OF THE

DIFFERENTIAL AND INTEGRAL

ELEMENTS

OF

ANALYTICAL GEOMETRY,

AND OF THE

DIFFERENTIAL AND INTEGRAL

CALCULUS.

BY

GERARDUS BEEKMAN DOCHARTY, LL.D.,

PROFESSOR OF MATHEMATICS IN THE NEW YORK FREE ACADEMY,
AND AUTHOR OF A PRACTICAL AND COMMERCIAL ARITHMETIC,
THE INSTITUTES OF ALGEBRA, AND THE ELE-
MENTS OF GEOMETRY AND TRIGONOMETRY.

NEW YORK:
HARPER & BROTHERS, PUBLISHERS,
FRANKLIN SQUARE.
1865.

Entered, according to Act of Congress, in the year one thousand eight hundred and sixty-five, by

HARPER & BROTHERS,

In the Clerk's Office of the District Court of the Southern District of New York.

TO

JAMES M. McLEAN, Esq.,

FOR THE

GREAT INTEREST HE HAS MANIFESTED IN THE CAUSE OF CORRECT EDUCATION, AND PARTICULARLY FOR HIS PERSONAL EFFORTS IN THE PROMOTION OF LITERATURE, SCIENCE, AND ART IN THE
NEW YORK FREE ACADEMY,

this volume is most respectfully dedicated

BY

THE AUTHOR.

PREFACE.

The following Treatise on the Elements of Analytical Geometry and the Differential and Integral Calculus is the result of long experience in educating young men. It is, in fact, the substance of that which the Author has taught in the Free Academy for several years; and one great object in bringing it before the scientific instructors of our country is an endeavor, on his part, to render the student's path, particularly through the Calculus, as smooth, and plain, and easy as the nature of that science will admit.

The Author well knows that a correct system of education requires constant mental effort on the part of the student who aspires to be a thorough scholar, and that a subject which is simplified too much in a text-book becomes unfitted for the object for which it was intended—namely, to enlarge and strengthen the powers of the mind. But he thinks that this fault can scarcely occur in any treatise on the "Theory of the Variation of Variables and their Functions," a science which, under the most favorable light that can be thrown upon it, is sufficiently obscure for the mind, at the early age in which the young men of this country begin the study of it.

It will be seen that the Author has used the old symbols exclusively, and that he has confined his system to the *Method of Limits*. He has done this, not on account of any partiality he feels for that system, for the *Infinitesimal Method* generally reaches the conclusions in a far less time and a much shorter space, but solely from his belief that the former is better adapted to the great purpose of teaching, and that it has a greater tendency to draw out the energies of the student, expand his genius, invigorate his thoughts, and impart a keener perception to his intellectual powers.

In discussing the Ellipse and Hyperbola together, the Author has followed the example of several eminent writers on Analytics. In this, as also in the method he has adopted in the discussion of the circle and the transformation of co-ordinates, and elsewhere, he has been enabled to reduce the Treatise to its minimum size without the omission of a single essential proposition.

In the recitation-room, perhaps it would be better for one student to take the proposition in reference to the ellipse, while another student takes the same for the hyperbola; in this way the number of examples becomes doubled.

With these prefatory remarks, the volume is respectfully submitted to the learned and scientific gentlemen who are engaged in teaching those useful and important branches of pure Mathematics. They will find it precisely what it purports to be—nothing more and nothing less—a Treatise on the *Elements* of Analytics and the Differential and Integral Calculus, prepared at the request, and for the use of the students in the New York Free Academy, but equally adapted to the instruction given in our colleges and other seminaries of learning.

May, 1865.

CONTENTS.

ANALYTICAL GEOMETRY.

CHAPTER I.

On the Point and Straight Line.—On the ——— ation of Co-ordinates .... Page 9–26

CHAPTER II.

On the Circle.—Examples.—Of the Parabola.—On the Ellipse and Hyperbola.—The Ellipse and Hyperbola referred to their Centre and Conjugate Diameters.—Problems.—General Equation of the Second Degree containing two Variables 26–81

CHAPTER III.

ANALYTICAL GEOMETRY OF THREE DIMENSIONS.

Equations of a Point in Space.—Distance between two Points in Space.—Equations of a ——— in Space ——— Plane ... 82–101

DIFFERENTIAL CALCULUS.

CHAPTER I.

Definitions and Introductory Remarks.—Functions defined.—Differentiation of Algebraic Functions.—Differential Coefficient of a Function of a Function.—Examples 103–130

CHAPTER II.

On successive Differential Coefficients 131–132
Taylor's Theorem 133–136
Maclaurin's Theorem 136–137
Development of Functions of two or more Variables when each receives an Increment 137–144
Transcendental Functions—Differential of the Logarithm of a Quantity 145
Examples 146–147
Differential of an Exponential Function 147–150
Circular Functions 150–159
Implicit Functions 159–161

CHAPTER III.

Application of the Calculus to the Theory of Curves....Page 161
Asymptotes 171–175
Singular Points of Curves... 175–185
The Logarithmic Curve 185
The Cycloid 186
Spirals 187
The Cissoid.. 189
Differentials of the Lengths and Area of Curves.............. 190–191
Differentials of Surfaces and Volumes of Revolution......... 191–193
Differentials of Areas, etc., referred to Polar Co-ordinates... 193–197
Asymptotes to Polar Curves....................................... 197–199
Osculatory Circle.—Radius of Curvature....................... 200–206
Radius of Curvature to Spirals 206–207
Evolutes and Involutes 208–215

CHAPTER IV.

Vanishing Fractions... 215–223
Maxima and Minima of Functions of one Variable 224–231
Maxima and Minima of Functions of two Variables 232–235
To change the Independent Variable 235–236
On Tangent Planes and Normal Lines to Curved Surfaces.. 236–238
Radius of Curvature of a Curve of double Curvature......... 238–239

ELEMENTS OF THE INTEGRAL CALCULUS.

Preliminary Remarks.—Integral of $x^n dx$...................... 240–243
Integration of Binomials 244–247
    "   of Circular Functions 247–252
    "   by Series... 252
    "   of Binomials... 253–256
    "   of Rational and Irrational Fractions............ 256–260
           by Parts... 260–265
    '   of Logarithmic Functions 265–269
    "   of Exponential Functions 269–271
    "   of Trignometrical Functions 271–272
    "   of Sines and Cosines of Multiple Arcs 273–274
    "   between Limits 274–275
Rectification and Quadrature of Plane Curves................... 275–284
Surfaces and Volumes of Revolution 284–292
Double Integrals... 292–295
The Volume a Function of three Variables 295–296
The Surface a Function of three Variables................... 296

The Theory of Infinitesimals...................................... 296–300

MISCELLANEOUS EXAMPLES.. 300–306

ANALYTICAL GEOMETRY.

PROBLEMS in Geometry, which, by the method of Euclid, would require for their solution long and laborious reasoning, and constant reference to propositions previously established, are often solved by analysis in a brief and easy manner, and from the most elementary principles. This is called *the application of Algebra to Geometry*.

Analytical Geometry has for its object a more general and extensive field than that of solving *determinate* problems by means of algebraical equations; it is more generally understood to be the analytical investigation of the general properties of geometrical magnitudes by means of *indeterminate* equations.

ANALYTICAL GEOMETRY OF TWO DIMENSIONS.

CHAPTER I.

(1.) *On the Point and straight Line.*

Assume the two straight lines AX, AY, whose position is known, situated in the same plane, and forming a given angle with each other.

Let P be any point in the same plane. We are required to determine its position.

From the point P draw the lines PC, PB respectively parallel to AX and AY. Then it is evident that the point P will be determined if we know the length of the lines CP and PB, or, which is the same thing, the lines AB and AC, of the parallelogram AP.

2. The two fixed lines AX, AY, extending indefinitely

right and left, up and down from it, are called the *axes*, or the *axes of co-ordinates*, AX being the axis of x, and AY the axis of y.

The distances AB or CP, and AC or BP, are called the *co-ordinates* of the point P; the former being the *abscissa*, and the latter the *ordinate*.

The variable abscissa is denoted by x, and any known abscissa by x', x'', x''', etc.

The variable ordinate is denoted by y, and any known ordinate by y', y'', y''', etc.

The point A is called the *origin of co-ordinates*.

The axes may be drawn making any angle with each other; when they form a right angle they are called *rectangular*, and when they do not form a right angle they are called *oblique axes*.

3. For every point situated on the axis of x we must have
$$y=0,$$
because that equation indicates that the distance of the point in question from that axis is nothing; likewise, for every point situated on the axis of y,
$$x=0.$$
Therefore, the system of the two equations
$$x=0, y=0,$$
characterizes the point A, the origin of co-ordinates, since these equations can hold good at the same time for no other point.

In general, the two equations
$$x=x', y=y',$$
when considered together, characterize a point situated at the distance x' from the axis of y, and at a distance y' from the axis of x.

The first of these equations, when considered separately, belongs to all the points of a straight line drawn parallel to the axis of y at a distance $AB=x'$; and the second, to all points of a straight line drawn parallel to the axis of x at a distance $AC=y'$.

Hence the system of the two equations together belongs to the point P, where these lines intersect, and *to this point alone*.

These expressions are the analytical representations of the point, and for this reason are called the *Equations of a Point*.

4. Not only the absolute values of the distances of the point from the two axes must be known, but the algebraic signs *plus* and *minus* must also be regarded. (Algebra, Art. 3.)

$x=x'$ and $y=y'$ are the equations of the point P.
$x=-x'$, $y=y'$ are the equations of the point P'.
$x=-x'$, $y=-y'$ are the equations of the point P''.
$x=x'$, $y=-y'$ are the equations of the point P'''.

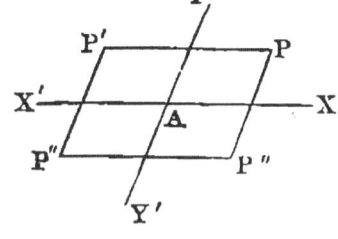

In the above we have considered the axes to be oblique to each. When the axes are at right angles, PB and PC are then perpendicular as well as parallel to the axes respectively, and the same remarks hold good in the one case as they do in the other.

PROPOSITION I.

(5.) *The distance between two given points in the plane of their co-ordinate axes is equal to the square root of the sum of the squares of the differences of their abscissas and ordinates.*

Let the co-ordinates of the point C, referred to the rectangular axes AX and AY, be x' and y', and the co-ordinates of the point B be x'', y'': it is required to determine the distance CB.

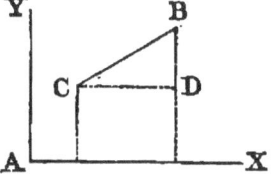

Draw CD parallel to AX; then
$$\overline{CB}^2 = \overline{CD}^2 + \overline{BD}^2.$$
But $CD = x'' - x'$, and $BD = y'' - y'$;
∴ $CB = \sqrt{(x''-x')^2 + (y''-y')^2}$ = the distance required.

This formula is perfectly general, and will apply to the case in which the two points are situated on different sides of the axes.

If one of the points, as C, be at the origin of co-ordinates, then $x'=0$, $y'=0$, and the formula becomes
$$D = \sqrt{x''^2 + y''^2}.$$

Examples.

1. Determine the point whose equations are
$x = -5, y = +4.$
2. Determine the point whose equations are
$x = +3, y = -6.$
3. Determine the point whose equations are
$x = -6, y = -5.$
4. Determine the point whose equations are
$x = 0, y = +7.$
5. Find the distance between two points whose co-ordinates are $x' = 3, y' = 6$; $x'' = 7, y'' = 9.$

Ans. 5.

6. Find the distance between the two points whose co-ordinates are $x' = +6, y' = +8$; $x'' = -6, y'' = -8.$

Ans. 20.

THE STRAIGHT LINE.

6. **Definition.** *The equation of a line is the equation which expresses the relation between the co-ordinates of every point of the line.*

PROPOSITION II.

7. *The equation of a straight line is*
$$y = ax + b,$$
where a is the ratio of the sines of the angles which the line makes with the co-ordinate axes when these axes form an oblique angle with each other, and a is the tangent of the angle which the line makes with the axis of x when the axes are at right angles. b is the distance from the origin to the point in which the line cuts the axis of y.

Let ECP be a straight line of indefinite length in a plane. Let A be the origin of co-ordinates, and AX, AY the axes inclined to each other at any given angle. Take *any* point P in the given line, and draw PB parallel to AY. Then will AB be the abscissa, and BP the ordinate of the point P. Through the origin A draw AD parallel to the line EP, meeting PB in D. Then will AB and BD be the co-ordinates of the point D.

Let the angle YAX be denoted by β, and the angle PEX=DAX be denoted by α.

Then, since PB is parallel to AY, and AD cuts them, the angle ADB is equal to YAD; that is, equal to $\beta-\alpha$.

Put $\quad$ AB=x, BD=y $\therefore$ BP=y+DP.

Then, by Trig.,

$$\frac{BD}{AB}=\frac{\sin.\alpha}{\sin.(\beta-\alpha)} \therefore \frac{y}{x}=\frac{\sin.\alpha}{\sin.(\beta-\alpha)}, \text{ or } y=\frac{\sin.\alpha}{\sin.(\beta-\alpha)}\cdot x, \quad (1)$$

which is the equation of the line AD passing through the origin.

If we now make DP, which is equal to AC=b, we shall have for the equation of the line EP,

$$y=\frac{\sin.\alpha}{\sin.(\beta-\alpha)}x+b. \quad (2)$$

Putting $\frac{\sin.\alpha}{\sin.(\beta-\alpha)}=a$, the equation of the line passing through the origin becomes

$$y=ax, \quad (3)$$

and the equation of the line cutting the axis of y above the origin becomes

$$y=ax+b. \quad (4)$$

If the line cut the axis of y below the origin, b would be negative. Hence, in the above equation, a and b may be any numbers, either whole or fractional, positive or negative.

If $\beta=90°$, the axes will be at right angles with each other, and equation (2) becomes

$$y=\frac{\sin.\alpha}{\sin.(90°-\alpha)}x+b$$
$$=\frac{\sin.\alpha}{\cos.\alpha}x+b$$
$$=\tan.\alpha.x+b. \quad \text{Put } \tan.\alpha=a.$$
$$\therefore y=ax+b,$$

in which a is the trigonometrical tangent of the angle E, radius being unity.

Illustration.

1. Let $\quad\quad\quad\quad y=ax-b.$
To construct it, make $\quad x=0.$
We have $\quad\quad\quad\quad y=-b,$
which is to be laid off from the origin A below the axis of x, and on the axis of y.

Let $y=0$. $\therefore x=\dfrac{b}{a}$,
which is to be laid off from the origin on the axis of x to the right, and the line joining those two points will be the line required.

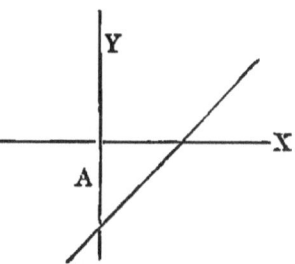

2. Let $y=-ax+b$.

Make $x=0$, then $y=+b$.

$y=0$, $x=+\dfrac{b}{a}$.

Lay off on the axis of y above the origin a distance $=b$, and on the axis of x to the right a distance $=\dfrac{b}{a}$, and join those points, as in the figure.

Let $y=-ax-b$.
Make $x=0$, then $y=-b$.

$y=0$, $x=-\dfrac{b}{a}$.

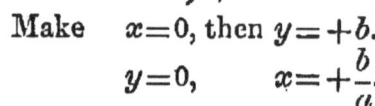

3. Let $y=ax$.
Make $x=0$, then $y=0$.
The line passes through the origin. Now, to find a second point, let $x=1$, then $y=a$.

Lay off on the axis of x to the right a distance equal to 1, and erect a perpendicular $=a$. Join the extremity of this perpendicular and the origin by a straight line, which will be the line required.

4. Let $y=b$.
Then the line is parallel to the axis of x.

Examples.

5. Construct the line whose equation is
$$y=2x+5.$$
6. Construct the line whose equation is
$$y=-3x+9.$$
7. Construct the line whose equation is
$$y=4x-3.$$
8. Construct the line whose equation is

$y = -5x - 8$.

9. Construct the line whose equation is
$y = 2x$.

10. Construct the line whose equation is
$y = -3x$.

PROPOSITION III.

(8.) *Every equation of the first degree containing two variables is the equation of a straight line.*

Every equation of the first degree containing two variables may be reduced to the form

$$Ay = Bx + C.$$
$$\therefore y = \frac{B}{A}x + \frac{C}{A}.$$

If we put $\frac{B}{A} = a$, and $\frac{C}{A} = b$, we shall have

$$y = ax + b.$$

When the axes are rectangular, $\frac{B}{A}$ is the tangent of the angle which the line makes with the axis of x; and $\frac{C}{A}$ is the distance from the origin to the point where the line cuts the axis of y, whatever angle the axes may make with each other.

If we solve the equation for x, we shall have

$$x = \frac{A}{B}y - \frac{C}{B},$$

where $\frac{A}{B}$ is the tangent of the angle which the line makes with the axis of y, and $-\frac{C}{B}$ is the distance from the origin to the point where the line cuts the axis of x.

Therefore, *if we determine the value of either of the variables, the co-efficient of the other variable will be the tangent of the angle which the line makes with the axis of that variable, and the absolute term will denote the distance from the origin to the point where the line cuts the axis of the variable by whose co-efficient we have divided the equation.*

Illustration.

If $3y = 6x - 9,$
then $y = 2x - 3.$

Here 2 is the tangent of the angle which the line makes with the axis of x, and -3 is the distance from the origin to the point where the line cuts the axis of y.

Or, solving the equation for x,

$$x = \frac{1}{2}y + \frac{3}{2}.$$

Here $\frac{1}{2}$ is the tangent of the angle which the line makes with the axis of y, and $+\frac{3}{2}$ is the distance from the origin to the point where the line cuts the axis of x.

(9.) *If the co-ordinates of any point of a line be substituted for the variables in the equation of that line, the equation will be satisfied.*

PROPOSITION IV.

(10.) *The equation of a straight line which passes through a given point, whose co-ordinates are x' and y', is*

$$y - y' = a(x - x').$$

Let P be the given point.
The equation of a straight line is of the form

$$y = ax + b. \qquad (1)$$

If we substitute the co-ordinates of the point P in this equation, we shall have

$$y' = ax' + b. \qquad (2)$$

Each of these equations must be satisfied at the same time; therefore, by subtracting the second from the first, we shall have

$$y - y' = a(x - x')$$

for the equation of a line passing through the given point.

Since the quantities x, y, and a are unknown, it follows that a, the tangent of the angle which the line makes with the axis, is indeterminate. There may, therefore, be an infinite number of straight lines drawn through a given point.

(11.) If it be required that the straight line, drawn through the given point, shall be parallel to a given line, then the angle which the line makes with the axis is known, and only one line can be drawn through a given point parallel to a given line.

Illustration.

If we were required to draw a line through the point P, whose abscissa AB is 8, and ordinate BP =3, parallel to the line AD, whose equation is

$$y = \frac{1}{2}x,$$

we shall have in the equation

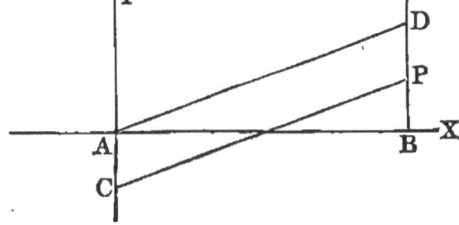

$$y - y' = a(x - x'),$$
$$y - 3 = \frac{1}{2}(x - 8),$$
$$2y - 6 = x - 8,$$
$$y = \frac{1}{2}x - 1.$$

Lay off from A a distance AC, equal to 1 below the axis of x on the axis of y, and on the axis of x, to the right of the origin, a distance equal to 2, and join those points by the line CP, which will be the line required.

PROPOSITION V.

(12.) *The equation of a straight line which passes through two given points whose co-ordinates are x', y' ; x'', y'', is*

$$y - y' = \frac{y'' - y'}{x'' - x'}(x - x').$$

Let C and B be two given points, the co-ordinates of C being x', y', and the co-ordinates of B being x'', y'', and x and y the co-ordinates of any other point whatever.

The general equation of the straight line is
$$y = ax + b. \tag{1}$$
But, since x', y', and x'', y'' are points in the straight line, these co-ordinates must satisfy the equation.
Hence
$$y' = ax' + b \tag{2}$$
$$y'' = ax'' + b. \tag{3}$$
Subtract (2) from (3), then
$$y'' - y' = a(x'' - x'). \tag{4}$$
From equation (4), $a = \dfrac{y'' - y'}{x'' - x'}$.

Again, subtract (2) from (1),
$$y - y' = a(x - x'). \tag{5}$$
Substitute in (5) the value of a obtained from (4), we have
$$y - y' = \frac{y'' - y'}{x'' - x'}(x - x'),$$
which is the equation of the line passing through the two given points C and B.

(13.) In the figure we see that CD is parallel to AX, and the line BC cuts them; therefore the angle BCD is equal to the angle which BC makes with the axis of x. Also that $y'' - y'$ is equal to DB, and $x'' - x'$ is equal to CD.

Hence $\quad \dfrac{y'' - y'}{x'' - x'} = \dfrac{DB}{CD} =$ tangent BCD, to radius unity.

If $\quad y'' = y'$,
the line would be parallel to the axis of x,
and $\quad \dfrac{y'' - y'}{x'' - x'} = 0$.

If $\quad x'' = x'$,
then $\quad \dfrac{y'' - y'}{x'' - x'} = \dfrac{y'' - y'}{0} = \infty$,
and the line would be perpendicular to the axis of x.

Examples.

1. Find the equation of a straight line which passes through the two points whose co-ordinates are $x' = 5$, $y' = 3$, and $x'' = 6$, $y'' = 2$, and determine the angle which it makes with the axis of x.

2. Find the equation of the straight line which passes

through the two points whose co-ordinates are $x'=3$, $y'=4$; $x''=5$, $y''=2$.

PROPOSITION VI.

(14.) *The tangent of the angle which two lines make by their intersection with each other is*

$$\frac{a'-a}{1+aa'},$$

where a and a' are the trigonometrical tangents of the angles which the two lines respectively make with the axis of abscissas, or the axis of x.

Let ED and CB be any two lines intersecting in P. Let the equation of the line ED be
$$y=ax+b,$$
and the equation of the line CB be $y=a'x+b'$.

Then a will be the tangent of the angle PEX, and a' the tangent of the angle PCX.

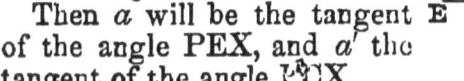

Let the angle PEX be represented by a, and the angle PCX be represented by a'.

Then, because PCX is the exterior angle of the triangle EPC, it is equal to the sum of the two interior opposite angles CPE and CEP. Put, therefore, V=EPC.
$$V=a'-a,$$
and $\quad \tan. V = \tan. (a'-a) = \dfrac{\tan. a' - \tan. a}{1+\tan. a, \tan. a'}.$

Hence $\quad \tan. V = \dfrac{a'-a}{1+aa'}.$

If the lines are parallel, they will not intersect,

and $\quad \tan. V = \dfrac{a'-a}{1+aa'} = 0. \therefore a'-a=0,$ or $a'=a.$

If the lines intersect at right angles,

then $\tan. V = \dfrac{a'-a}{1+aa'} = \infty. \therefore 1+a'a=0,$ or $a'=-\dfrac{1}{a}$

which is the equation of condition for two lines to be at right angles with each other.

(15.) Hence it appears that if the equation to a straight line be $y=ax+b$, the equation to a straight line perpendicular to it, and passing through a given point whose co-ordinates are x', y', will be

$$y-y'=-\frac{1}{a}(x-x').$$

Proposition VII.

(16.) *The length of a straight line drawn from a given point perpendicular to a given straight line is*

$$\frac{y'-ax'-b}{\sqrt{1+a^2}},$$

where x' and y' are the co-ordinates of the given point.

Let EB be the given line, and P the given point. The equation of the given line is

$$y=ax+b. \qquad (1)$$

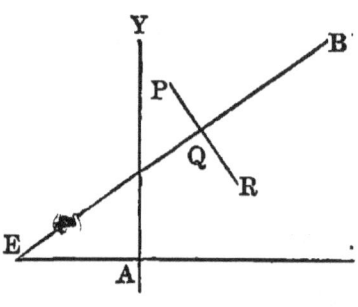

Then, since PQR is drawn through a point whose co-ordinates are x' and y', and perpendicular to a straight line whose equation is $y=ax+b$, the equation to PR is, by the last proposition (Art. 15),

$$y-y'=-\frac{1}{a}(x-x'). \qquad (2)$$

Now, to find the length of PQ, we must find the co-ordinates of the point Q, in which PR meets EB, and then substitute x', y', and the co-ordinates of Q in the general expression for the distance between two points, viz.:

$$D=\sqrt{(x'-x'')^2+(y'-y'')^2}. \qquad (3)$$

Let us call the co-ordinates of the point Q x'' and y''. Then, since Q is a point in the straight line EB, the co-ordinates of that point must satisfy equation (1).

$$\therefore y''=ax''+b.$$

For the sake of convenience, we will put this equation under the form

$$y''-y'=a(x''-x')-y'+ax'+b, \qquad (4)$$

which is done by subtracting y' from each member of the equation, and adding ax' to, and subtracting it from the second member.

But, since Q is a point in the straight line PR, its coordinates x'', y'' must satisfy equation (2).

$$\therefore y'' - y' = -\frac{1}{a}(x'' - x'). \qquad (5)$$

Now equations (4) and (5) must both hold good for the point Q. Therefore, subtracting (5) from (4), we have

$$0 = \frac{1}{a}(x'' - x') + a(x'' - x') - y' + ax' + b.$$

$$\therefore x'' - x' = a\frac{y' - ax' - b}{1 + a^2}.$$

Substitute this value of $x'' - x'$ in equation (5), we shall have

$$y'' - y' = -\frac{y' - ax' - b}{1 + a^2}.$$

Substituting the values of $x'' - x'$ and $y'' - y'$ in equation (3), we have

$$D = \sqrt{a^2\left(\frac{y' - ax' - b}{1 + a^2}\right)^2 + \left(\frac{y' - ax' - b}{1 + a^2}\right)^2}.$$

$$\therefore D = \frac{y' - ax' - b}{\sqrt{1 + a^2}},$$

which is the length of the perpendicular required.

ON THE TRANSFORMATION OF CO-ORDINATES.

(17.) When a line is represented by an equation in reference to any system of axes, we can always transform that equation into another in reference to a new system of axes, chosen at pleasure, which equation shall equally express the relations between the co-ordinates of every point of the line.

This is called *the transformation of co-ordinates*. We may either change the origin without altering the relative position of the axes, or alter the relative position of the axes without changing the origin; or we may change the position of the origin and the angle of inclination of the axes to each other at the same time.

Proposition VIII.

(18.) *The formulas for passing from one system of co-ordinate axes to another system respectively parallel to the first are*

$$x = a + x_{,},$$
$$y = b + y_{,}.$$

Let AX, AY be the original axes; and A'X', A'Y' be the new axes parallel to the former.

Let $x = $ AM, the abscissa of the point P, referred to the original axis, and $y = $ MP, the ordinate, referred to the same.

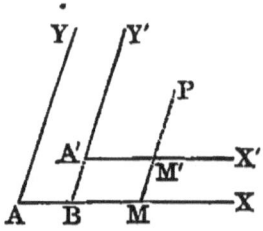

Let $x_{,} = $ A'M' $= $ BM, the abscissa referred to the new axis; and $y_{,} = $ M'P $= $ the ordinate.

Let the co-ordinates of the new origin AB, BA' be represented by a and b. Then we shall have

AM $=$ AB$+$BM, or $x = a + x_{,}$,
and   MP $=$ MM'$+$M'P,
or   MP $=$ BA'$+$M'P, or $y = b + y_{,}$,

which are the formulas required.

Proposition IX.

(19.) *The formulas for passing from a system of oblique co-ordinates to another system also oblique are,*

$$x = \frac{x_{,} \sin.(\beta - \alpha) + y_{,} \sin.(\beta - \alpha')}{\sin. \beta},$$

$$y = \frac{x_{,} \sin. \alpha + y_{,} \sin. \alpha'}{\sin. \beta},$$

where β is the angle contained between the original axes, and α and α' are the inclination of the new axes to the original axis of abscissas.

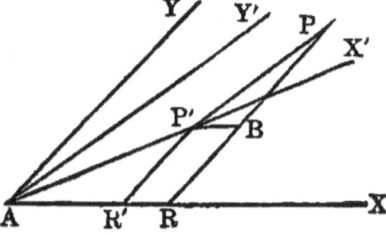

Let AX, AY be the primitive axes; AX', AY' the new axes; and P any point in the plane.

Let the co-ordinates of the point P referred to the

ON THE TRANSFORMATION OF CO-ORDINATES. 23

primitive axes be denoted by x and y; and its co-ordinates referred to the new axes be $x_{,}$ and $y_{,}$.

Through the point P draw PP' parallel to AY' and PR parallel to AY; also, through P' draw P'B parallel to AX, and P'R' parallel to AY.

Denote the angle YAX by β, the angle X'AX by α, and the angle Y'AX by α'.

Then, by Trigonometry,

$\dfrac{AR'}{AP'} = \dfrac{\sin.(\beta-\alpha)}{\sin.\beta} \therefore AR' = AP' \dfrac{\sin.(\beta-\alpha)}{\sin.\beta} = x_{,}\dfrac{\sin.(\beta-\alpha)}{\sin.\beta}$,

$\dfrac{P'B}{P'P} = \dfrac{\sin.(\beta-\alpha')}{\sin.\beta} \therefore P'B = P'P\dfrac{\sin.(\beta-\alpha')}{\sin.\beta} = y_{,}\dfrac{\sin.(\beta-\alpha')}{\sin.\beta}$,

$\dfrac{P'R'}{AP'} = \dfrac{\sin.\alpha}{\sin.\beta} \therefore P'R' = AP'\dfrac{\sin.\alpha}{\sin.\beta} = x_{,}\dfrac{\sin.\alpha}{\sin.\beta}$,

$\dfrac{BP}{P'P} = \dfrac{\sin.\alpha'}{\sin.\beta} \therefore BP = PP'\dfrac{\sin.\alpha'}{\sin.\beta} = y_{,}\dfrac{\sin.\alpha'}{\sin.\beta}$.

But $x = AR' + R'R = \dfrac{x_{,}\sin.(\beta-\alpha) + y_{,}\sin.(\beta-\alpha')}{\sin.\beta}$,

" $y = RB + BP = \dfrac{x_{,}\sin.\alpha + y_{,}\sin.\alpha'}{\sin.\beta}$,

the formulas required.

Cor. 1. If $\beta = 90°$, the primitive axes are at right angles with each other, and the formulas become

(20.) $x = x_{,}\cos.\alpha + y_{,}\cos.\alpha'$,
$y = x_{,}\sin.\alpha + y_{,}\sin.\alpha'$,

which are *the formulas for passing from a system of rectangular to a system of oblique co-ordinates, the origin remaining the same.*

Cor. 2. If $\alpha' - \alpha = 90°$, or $\alpha' = 90° + \alpha$, then cos. $\alpha' =$ cos. $(90° + \alpha) = -\sin.\alpha$ (Trig., Eq. 12), and sin. $\alpha' = \sin.(90°+\alpha) = \cos.\alpha$. Substituting these in the above formulas for cos. α' and sin. α', we have

(21.) $x = x_{,}\cos.\alpha - y_{,}\sin.\alpha$,
$y = x_{,}\sin.\alpha + y_{,}\cos.\alpha$,

which are the *formulas for passing from a system of rectangular co-ordinates to another system also rectangular, the origin remaining the same.*

If the position of the origin is changed also, then (20) becomes

(22.) $x = a + x_{,}\cos.\alpha + y_{,}\cos.\alpha'$,
$y = b + x_{,}\sin.\alpha + y_{,}\sin.\alpha'$.

And (21) becomes
(23.)
$$x = a + x_{,} \cos. \alpha - y_{,} \sin. \alpha,$$
$$y = b + x_{,} \sin. \alpha + y_{,} \cos. \alpha.$$

POLAR CO-ORDINATES.

(24.) In the preceding propositions we have determined the position of a line upon a plane by means of an equation between two variables expressing the distances of each of its points from two fixed straight lines, the distances being reckoned parallel to each of these lines respectively. There is, however, another method for determining a series of points, which, in certain cases, is more convenient.

To explain this mode of representing lines analytically, let us take any point, P, and let A'D be a given straight line in the plane of the curve, and A' a given point in that line. From A' draw the straight line A'P. Put A'P = r, and the angle PA'D = θ.

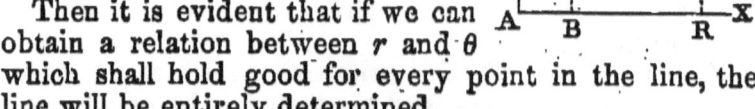

Then it is evident that if we can obtain a relation between r and θ which shall hold good for every point in the line, the line will be entirely determined.

The variable quantities r and θ are called *Polar Co-ordinates*.

The *point* A' is called *the pole*, r *the radius vector*, and the relation between r and θ is termed the *Polar Equation*.

The Polar Equation is sometimes determined at once from some known property; it is, however, more usually deduced from an equation between rectilinear co-ordinates, which may easily be effected by a transformation.

PROPOSITION X.

(25.) *The formulas for passing from a system of rectangular to a system of polar co-ordinates are*
$$x = a + r \cos. \theta,$$
$$y = b + r \sin. \theta,$$
r being the radius vector, and θ the variable angle which it makes with the axis of x.

Let AX and AY be the rectangular axes, A' the pole, and A'D, parallel to AX, be the line from which the variable angle is to be estimated.

Put $PA'D = \theta$,
$A'P = r$.

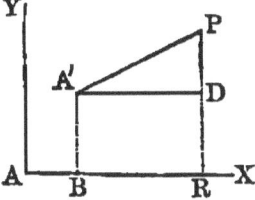

Let the co-ordinates of the point P, referred to the primitive axes, be x and y, and the co-ordinates of A', referred to the same, be a and b.

Then $x = AB + BR$,
and $y = RD + DP$.
But $BR = A'D = A'P \cos. PA'D = r \cos. \theta$,
and $DP = A'P \sin. PA'D = r \sin. \theta$.
Hence $x = a + r \cos. \theta$,
$y = b + r \sin. \theta$.

26. SCHOLIUM. If A' be placed at the origin A, then $a = 0$, $b = 0$, and the equations become
$x = r \cos. \theta$,
$y = r \sin. \theta$.

Problem.

(27.) *To find the co-ordinates of the point of intersection of two lines.*

Let the equation of the line ED be $y = ax + b$,
and the equation of the line BC be $y = a'x + b'$.

To find the abscissa and ordinate of the point P, their intersection.

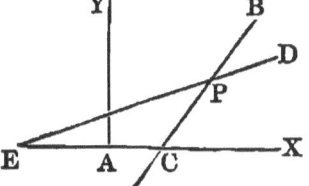

If we call the co-ordinates of the point P x, y, then, because that point is common to both lines, the co-ordinates of that point must satisfy both equations; hence, eliminating y, we shall have
$$x = \frac{b' - b}{a - a'}.$$

Hence $$y = \frac{ab' - a'b}{a - a'}.$$

SCHOLIUM. Now if $a = a'$, the two lines become parallel, and will never meet. This is evident from the anal-

B

ysis also; for $a=a'$ renders both x and y equal to infinity.

If we have, respectively, the equations of any two lines whatever on the same plane, referred to the same axes, we can find the point or points of intersection simply by finding the values of x and y from those equations. As these values are co-ordinates to both lines, they must be the values of the co-ordinates of the point of intersection.

CHAPTER II.

ON THE CIRCLE.

(28.) The circumference of the circle is a plane curve, every point of which is equally distant from a point within called the centre.

PROPOSITION I.

(29.) *The equation of the circle referred to oblique axes is*

$$(x-x')^2 + (y-y')^2 \pm 2(x-x')(y-y') \cos. \beta = R^2,$$

where R is the radius of the circle, x and y are the co-ordinates of any point in the circumference, x', y' the co-ordinates of the centre, and β the inclination of the axes.

Let P be any point in the circumference of the circle whose centre is C.

Draw AY, AX, the axes of co-ordinates, making an angle equal to β. Through C and P draw CB and PE both parallel to AY, and through C draw FCG parallel to AX. 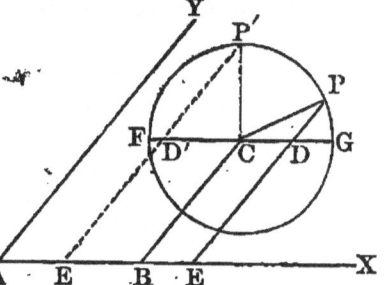 Let the co-ordinates of the centre, AB, BC, be denoted by x', y', and AE, EP, the co-ordinates of the point P, by x, y. Then, because FG is parallel to AX, and PE is parallel to AY, the angle PDG is equal to YAX equal to β; and therefore PDC is the supplement of PDG$=180°-\beta$. But by Plane Trig., Th. V.,

ON THE CIRCLE. 27

$$\cos. \text{PDC} = \frac{\text{CD}^2 + \text{DP}^2 - \text{CP}^2}{2\ \text{CD} \times \text{DP}} = \frac{(x-x')^2 + (y-y')^2 - \text{R}^2}{2(x-x')(y-y')};$$

$$\therefore \cos. \text{PDC} = -\cos. \beta = \frac{(x-x')^2 + (y-y')^2 - \text{R}^2}{2(x-x')(y-y')};$$

$$\therefore (x-x')^2 + (y-y')^2 \pm 2(x-x')(y-y')\cos.\beta = \text{R}^2, \quad (1)$$

the equation required.

(30.) SCHOLIUM 1. If $\beta = 90°$, the axes are rectangular, and (Eq. 1) becomes

$$(x-x')^2 + (y-y')^2 = \text{R}^2, \quad (2)$$

which is *the most general equation of the circle referred to rectangular axes.*

SCHOLIUM 2. If the origin is removed from A to F, and the axes are rectangular,

x' becomes equal to R, and

$$y' = 0,$$

and the equation becomes

$$y^2 = 2\text{R}x - x^2, \quad (3)$$

the equation of the circle when the origin of co-ordinates is on the circumference.

(31.) SCHOLIUM 3. If the origin is transferred to the centre, then $x' = 0$, $y' = 0$, and the equation is

$$y^2 + x^2 = \text{R}^2, \quad (4)$$

which is *the equation of the circle when the origin of co-ordinates is at the centre.*

(32.) If we wish to find the points in which the curve cuts the axes, first make

$$x = 0,$$

and we shall have, from (Eq. 4),

$$y = \pm \text{R},$$

which shows that the curve cuts the axis of y in two points on different sides of the origin at a distance equal to the radius.

If we make $y = 0$, then

$$x = \pm \text{R},$$

which shows that the curve cuts the axis of x at a distance equal R on the right and R on the left of the origin.

(33.) If we wish to trace the curve through the intermediate points, we find the value of x or y thus:

$$x = \pm \sqrt{\text{R}^2 - y^2},$$

or
$$y = \pm \sqrt{\text{R}^2 - x^2}.$$

Now, since every value of one variable gives two equal values of the other, with contrary signs, it follows that the curve is symmetrical in regard to both axes.

If we make x or y greater than R, the value of the other becomes imaginary, which shows that the curve does not extend on either side of the origin beyond the value of x or $y = \pm R$.

In a manner entirely similar, if in (Eq. 3) we make
$$y = 0,$$
we have $\quad 0 = 2Rx - x^2$, or $(2R - x)x = 0$,
whence $\quad\quad x = 0$, and $x = 2R$,
which shows that the curve passes through the origin of co-ordinates, and also cuts the axis of x at a distance from the origin equal to twice the radius.

PROPOSITION II.

(34.) *The equation of a tangent line to the circle is*
$$yy'' + xx'' = R^2,$$
where R is the radius of the circle, and x'', y'' are the co-ordinates of the point of tangency, and x, y are the general co-ordinates of the tangent line.

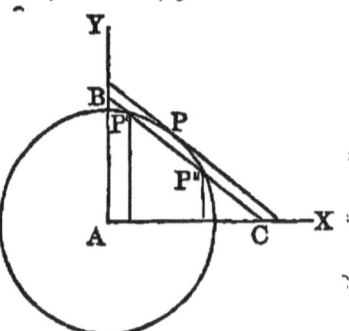

Let the co-ordinates of the point P' be denoted by x', y', and the co-ordinates of P'' by x'', y''; then the equation of the line BC, passing through these two points, will be

$$y - y'' = \frac{y'' - y'}{x'' - x'}(x - x''). \quad (1)$$

But, since these two points are on the circumference, their co-ordinates will satisfy the equation of the curve. Therefore
$$x''^2 + y''^2 = R^2, \quad\quad (2)$$
$$x'^2 + y'^2 = R^2. \quad\quad (3)$$
Subtracting (3) from (2), we have
$$x''^2 - x'^2 + y''^2 - y'^2 = 0.$$
Factor these, we have
$$(x'' + x')(x'' - x') + (y'' + y')(y'' - y') = 0;$$

ON THE CIRCLE.

$$\therefore \frac{y''-y'}{x''-x'} = -\frac{x''+x'}{y''+y'}.$$

Substitute this value in equation (1), we have

$$y-y'' = -\frac{x''+x'}{y''+y'}(x-x''). \qquad (4)$$

Now, as the secant line BC moves toward P, the points P'' and P' approach each other, and ultimately coalesce at the point P; then the secant becomes a tangent, and x' becomes equal to x'' and y' equal to y'', and equation (4) becomes

$$y-y'' = -\frac{x''}{y''}(x-x''),$$

which is the equation of the tangent line. This may be reduced to a more simple form.
Clearing fractions, $yy''-y''^2 = -xx''+x''^2$,
or $\qquad yy''+xx'' = x''^2+y''^2$;
but $\qquad x''^2+y''^2 = R^2$;
$\qquad \therefore yy''+xx'' = R^2$,
or $\qquad y = -\frac{x''}{y''}x + \frac{R^2}{y''}.$

By comparing this with the equation of a right line, we see that $-\frac{x''}{y''}$ is the tangent of the angle which the tangent line makes with the axis of X, and $\frac{R^2}{y''}$ is the distance from the centre to the point where the tangent line cuts the axis of Y.

(35.) Definition. *A normal is a line perpendicular to the tangent at the point of contact, and limited by the axis of abscissas.*

PROPOSITION III.

(36.) *Every normal line in a circle passes through the centre.*
The equation of the tangent line is

$$y-y'' = -\frac{x''}{y''}(x-x'');$$

and as the normal line is perpendicular to the tangent, we have, from (Art. 15),

$$y-y''=\frac{y''}{x''}(x-x'').$$

Reducing, $yx''-x''y''=xy''-x''y'';$

$$\therefore y=\frac{y''}{x''}x.$$

Since this has no absolute term, it is the equation of a line passing through the origin; and as the origin is at the centre of the circle, the normal passes through the centre.

(37.) Definition. *Two lines which are drawn from the extremities of any diameter of a curve, and which intersect the curve at the same point, are called supplementary chords.*

Proposition IV.

(38.) *The supplementary chords in a circle are perpendicular to each other.*

Let A be the origin of co-ordinates, and A, B the extremities of a diameter through which the supplementary chords AP, BP are drawn. We are required to prove that these chords are perpendicular to each other.

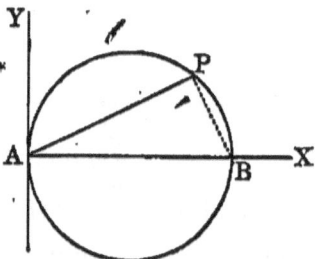

The equation of a straight line passing through a given point is of the form

$$y-y'=a(x-x'). \qquad (1)$$

If the line pass through A, the co-ordinates of which are $x'=0$, $y'=0$, the equation becomes

$$y=ax. \qquad (2)$$

If the line pass through B, whose co-ordinates are $x'=2r$, $y'=0$, the equation becomes

$$y=a'(x-2r). \qquad (3)$$

Now, if these two lines intersect each other, the co-ordinates of the point of intersection will satisfy (2) and (3). Hence, multiplying (2) and (3) together, the resulting equation,

$$y^2=aa'(x^2-2rx), \qquad (4)$$

will express the condition that the two straight lines shall intersect on the plane of the co-ordinate axes. But

if the point of intersection is to be in the circumference of a circle, x and y must satisfy the equation
$$y^2 = 2rx - x^2,$$
$$= -(x^2 - 2rx). \qquad (5)$$
Equating these two values of y^2, we have
$$aa'(x^2 - 2rx) = -(x^2 - 2rx);$$
$$\therefore aa' = -1,$$
which is the equation of condition for two lines to be perpendicular to each other. Hence the supplementary chords in a circle are at right angles to each other.

PROPOSITION V.

(39.) *The polar equation of the circle when the pole is on the circumference is*
$$r = \pm 2R \cos. \theta,$$
where R denotes the radius of the circle, r the radius vector, and θ the variable angle.

The equation of the circle referred to rectangular axes when the origin is at the centre, Art. 28, is
$$y^2 + x^2 = R^2. \qquad (1)$$
And the formulas for passing from a system of rectangular to polar co-ordinates, Art. 22, are
$$x = a + r \cos. \theta,$$
$$y = b + r \sin. \theta.$$
Squaring these equations, and substituting the values of x^2 and y^2 in (1), we have
$$b^2 + 2b.r \sin. \theta + r^2 \sin.^2 \theta + a^2 + 2ar \cos. \theta + r^2 \cos.^2 \theta = R^2;$$
or, arranging according to the powers of r,
$$r^2(\sin.^2 \theta + \cos.^2 \theta) + 2(b \sin. \theta + a \cos. \theta)r + a^2 + b^2$$
$$= R^2, \qquad (2)$$
which is the general polar equation of the circle.

If we place the pole at the right extremity of the diameter on the axis of x,
$$a = +R, \text{ and } b = 0.$$
The equation (2) reduces to $r^2 + 2Rr \cos. \theta = 0$, because $\sin.^2 \theta + \cos.^2 \theta = 1$;
$$\therefore r = -2R \cos. \theta. \qquad (3)$$
If we place the pole at the extremity of the diameter to the left of the centre, $a = -R$, and $b = 0$; and (2) reduces to $r^2 - 2Rr \cos. \theta = 0$;
$$\therefore r = 2R \cos. \theta. \qquad (4)$$

In (4), when $\theta=0$, cos. $\theta=1$; $\therefore r=2$ R.

As θ increases from 0 to 90°, the radius vector determines all the points in the upper semicircumference; and when $\theta=90°$, cos. $\theta=0$. From 90° to 270° the radius will be negative, and will determine no point of the circumference. From 270° to 360° the radius vector will again be positive, and determine all the points in the lower semicircumference.

The student may take equation (3) and illustrate the manner in which the radius vector will determine all the points of the circumference.

Examples.

1. Construct the line whose equation is
$$y=\frac{3}{5}x+\frac{2}{5}.$$

2. Construct the line whose equation is
$$y=\frac{3}{4}x-\frac{5}{8}.$$

3. Construct the line whose equation is
$$y=-3x+6.$$

4. Construct the lines whose equation is
$$y^2-5x=0.$$

5. Construct the lines whose equation is
$$y^2=7y-12.$$

6. Find the equation of the straight line which passes through two points whose co-ordinates are $x'=1$, $y'=3$, $x''=4$, $y''=5$. *Ans.* $y=x+1$.

7. Describe the circle whose equation is
$$y^2+x^2+4y-4x=8.$$

8. Describe the circle whose equation is
$$y^2+x^2-6y+8x=11.$$

9. Prove that the straight lines drawn from the angular points of a triangle to bisect the sides, pass through the same point.

Let the lines CF, BE, and AD bisect the opposite sides of the triangle ABC; we are

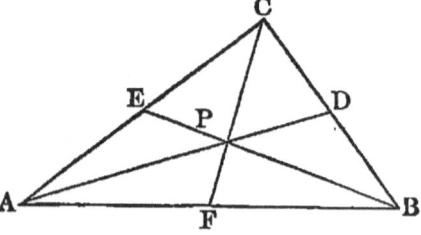

required to prove that they pass through the same point, P.

Let FB, FC be the axes of co-ordinates, the origin being at F.

Then the co-ordinates of the point C are $x'=0$, and y';
" " " " B are x' and $y'=0$;
and, consequently, $y'=0$ and $-x'$ will be the co-ordinates of the point A.

But, since AE is half of AC, the co-ordinates of E will be $-\frac{1}{2}x'$ and $\frac{1}{2}y'$, and the co-ordinates of D, for a like reason, will be $\frac{1}{2}x$ and $\frac{1}{2}y'$.

Therefore the equation of the line AD, which is of the form
$$y-y'=\frac{y''-y'}{x''-x'}(x-x'),$$
becomes
$$y=\frac{\frac{1}{2}y'}{\frac{1}{2}x'+x'}(x+x'). \qquad (1)$$
And the equation of BE becomes
$$y=-\frac{\frac{1}{2}y'}{\frac{1}{2}x'+x'}(x-x'). \qquad (2)$$

Now, where AD cuts CF, $x=0$,
and where BE cuts CF, $x=0$.
Making these substitutions, each equation gives
$$y=\tfrac{1}{3}y'.$$
Hence each of the lines intersects the line CF in the same point P.

10. Prove that the perpendiculars drawn from the angular points of a triangle to the opposite sides pass through the same point.

Let the perpendiculars AD, BD, and CF be drawn from the angular points A, B, and C to the sides of the triangle; we are required to prove that they pass through the same point P.

Let AX and AY be the rectangular axes. Now if it can be proved that AF is the abscissa of the point of intersection of the perpendiculars AD, BE, as well as of the point C, the truth of the problem will be established.

Let the co-ordinates of the point C be x', y'.
" " " " B be x'', $0=y''$.

Because AC passes through the origin and through the given point C, its equation is
$$y = \frac{y'}{x'}x.$$
The equation of the line BC, passing through two points whose co-ordinates are x'' and $y''=0$, will be
$$y = \frac{y'}{x'-x''}(x-x'').$$
The equation of the perpendicular AD is
$$y = -\frac{x'-x''}{y'}x.$$
The equation of the perpendicular BE is
$$y = -\frac{x'}{y'}(x-x'').$$
And at the point P, where these two perpendiculars intersect, the ordinates must be identical.
$$\therefore \frac{x'}{y'}(x-x'') = \frac{x'-x''}{y'}x;$$
$$\therefore x = x''.$$
That is, the abscissa of the point of intersection of AD and BE is the same as the abscissa of the point C; hence the perpendicular CF passes through the intersection P.

On the Parabola.

(40.) *Definition. A parabola is a plane curve, every point of which is equally distant from a fixed point and a given straight line.*

The fixed point is called the *focus* of the parabola, and the given straight line is called the *directrix*.

Proposition I.

(41.) *The equation of the parabola referred to rectangular axes whose origin is at the vertex of the axis is*
$$y^2 = 2px,$$
where x and y are the general co-ordinates of the curve, and $2p$ is the parameter of the axis.

Let F be the given fixed point, and DC the given straight line.

Draw FB perpendicular to DC, and bisect BF in A; then, by the definition, A is a point in the parabola. Take any other point, P, in the curve, and join PF. From P draw PD perpendicular to DC. From A draw AY perpendicular to AFX. A will then be the origin, and AX, AY the axes of coordinates.

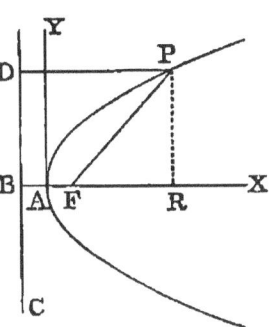

From P let fall the perpendicular PR on AX. Put $AR=x$, $RP=y$, $BF=p$; then $AF=\tfrac{1}{2}p$. Then, by the definition of the curve,
$$PF = DP = AR + AB = x + \tfrac{1}{2}p,$$
and
$$FR = x - \tfrac{1}{2}p.$$
But $\overline{PR}^2 + \overline{FR}^2 = \overline{PF}^2$, or $y^2 + (x-\tfrac{1}{2}p)^2 = (x+\tfrac{1}{2}p)^2$;
$$\therefore y^2 = 2px,$$
which is the equation of the parabola.

In order to find the value of the ordinate passing through the focus, make $x = \tfrac{1}{2}p$; then
$$y^2 = p^2;$$
$$\therefore y = \pm p,$$
which shows that the double ordinate passing through the focus is equal to the parameter.

Solving the equation for y, we have
$$y = \pm \sqrt{2px},$$
which shows that the curve is symmetrical with respect to the axis of x; and for all values of x negative, the value of y is imaginary; hence there is no part of the curve to the left of the origin A.

If $x=0$, $y=0$, and the curve passes through the origin.

If we give a succession of values for x, we perceive that as x increases, y increases also, and that for each value of x there will be two values of y numerically equal with opposite signs. Hence the curve extends indefinitely to the right of A.

If we put the equation into a proportion, we have
$$x : y :: y : 2p.$$

(42.) That is, *the parameter of the axis is a third proportional to any abscissa and its corresponding ordinate.*

(43.) If we take any two points on the curve, and designate their co-ordinates by x', y', and x'', y'', we shall have
$$y'^2 = 2px',$$
$$y''^2 = 2px''.$$
Dividing these equations member by member, we shall have
$$\frac{y'^2}{y''^2} = \frac{x'}{x''},$$
or
$$y'^2 : y''^2 :: x' : x''.$$
(44.) That is, *the squares of the ordinates are to each other as their corresponding abscissas.*

Proposition II.

(45.) *The equation of a tangent line to the parabola is*
$$yy'' = p(x + x''),$$
where p denotes half the parameter, and x'', y'' are the co-ordinates of the point of tangency.

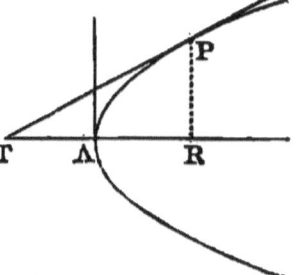

Put $x'' = AR$, $y'' = RP$; then the equation of a line passing through P will be
$$y - y'' = a(x - x'').$$
But since the point P is on the curve, its co-ordinates must satisfy the equation of the curve.
(1) $\because y^2 = 2px$ is the equation of the parabola where x and y denote the co-ordinates of any point of the curve.
(2) $\therefore y''^2 = 2px''.$
Subtracting (2) from (1), we have
$$y^2 - y''^2 = 2p(x - x''),$$
or $(y + y'')(y - y'') = 2p(x - x'').$
Substituting for $y - y''$ its value as above, we have
$$(y + y'')a(x - x'') = 2p(x - x'').$$
(3) $\therefore a = \dfrac{2p}{y + y''}.$

But, in order that the line TP shall be tangent to the curve at P, x and y, the general co-ordinates, must, in this case, become the co-ordinates of the point P.
$$\therefore x = x'' \text{ and } y = y''.$$

ON THE PARABOLA. 37

$$\therefore a = \frac{2p}{2y''} = \frac{p}{y''}.$$

(4) $\therefore y - y'' = \frac{p}{y''}(x - x'')$ is the equation of the tangent line. This, however, can be reduced.

By clearing the equation of fractions, we have
$$yy'' - y''^2 = px - px'';$$
but
$$y''^2 = 2px''.$$
Substitute this, $yy'' - 2px'' = px - px''$,
$$yy'' = p(x + x'').$$

To find the point at which the tangent cuts the axis of abscissas, make $y = 0$, and we have
$$p(x + x'') = 0,$$
or
$$x = -x'',$$
where $x =$ AT, the distance from the origin to the point in which the tangent line cuts the axis of X.

(46.) Definition. *The subtangent is the projection of the tangent on the axis of abscissas.*

The subtangent TR is therefore bisected at the vertex of the parabola.

Hence we are enabled to draw a tangent at any point of the curve, thus:

From a given point let fall a perpendicular on the axis, and from the origin lay off, on the axis, to the left a distance equal to the distance from the origin to the foot of the perpendicular. Join this last point and the given point by a straight line, which will be the tangent required.

(47.) Definition. *A subnormal is the projection of the normal on the axis of abscissas.*

PROPOSITION III.

(48.) *The equation of a normal line to the parabola is*

$$y - y'' = -\frac{y''}{p}(x - x''),$$

where x'' and y'' are the co-ordinates of the point of tangency.

For the equation of a line passing through P is of the form
$$y - y'' = a'(x - x''); \qquad (1)$$

and since the normal PN is
perpendicular to the tangent
TP (Art. 34), we must have

$$a' = -\frac{1}{a}.$$

But by Prop. II., $a = \dfrac{p}{y''}$;

$$\therefore a' = -\frac{y''}{p}.$$

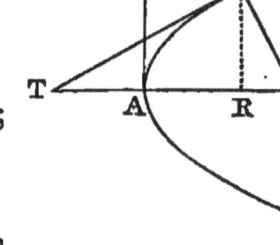

Substituting this in (Eq. 1),
we have $\quad y - y'' = -\dfrac{y''}{p}(x - x'').\qquad (2)$

49. To find the point in which the normal cuts the axis of X, make $y = 0$ in (Eq. 2), and we have
$$x - x'' = p.$$
Here $x = $ AN and $x'' = $ AR; $\therefore x - x'' = $ RN = subnormal $= p$. Hence *the subnormal is constant, and equal to half the parameter.*

(50.) This furnishes us with a simple method of drawing a tangent to any point of a parabola. Thus: From the given point let fall a perpendicular on the axis, and from the foot of the perpendicular lay off a distance equal to half the parameter, on the axis, to the right; join this and the given point by a straight line, and at the given point draw a perpendicular to this latter line, which will be the tangent required.

Proposition IV.

(51.) *A tangent to the parabola at any point makes equal angles with the axis and with the line drawn from the point of tangency to the focus.*

Let TP be a tangent to a parabola, PF the line drawn from the point of tangency to the focus F. Then will the angle TPF be equal to the angle PTF. Designate the co-ordinates of the point P by x'' and y'', and the co-ordinates of the focus by x' and y'.

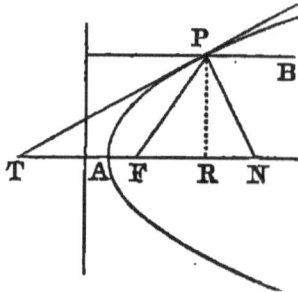

The equation of a line passing through the focus F will be of the form.
$$y - y' = a'(x - x'); \quad (1)$$
and if it passes through the point of tangency P, we shall have
$$y'' - y' = a'(x'' - x'). \quad (2)$$
But the co-ordinates of the focus give
$$x' = \tfrac{1}{2}p, \text{ and } y' = 0.$$
Therefore Eq. 2 becomes
$$y'' = a'(x'' - \tfrac{1}{2}p). \quad \therefore a' = \frac{y''}{x'' - \tfrac{1}{2}p}.$$

Now the tangent of the angle which the two lines TP and FP make at their intersection, by Art. 14, gives
$$\tan. \text{TPF} = \frac{a' - a}{1 + a'a}.$$
Substitute for a' its value $\dfrac{y''}{x'' - \tfrac{1}{2}p}$, and for a its value $\dfrac{p}{y''}$, obtained in Prop. II., and we shall have
$$\tan. \text{TPF} = \left(\frac{y''}{x'' - \tfrac{1}{2}p} - \frac{p}{y''} \right) \div \left(1 + \frac{p}{x'' - \tfrac{1}{2}p} \right);$$
which, by reduction, putting $y''^2 = 2px''$, becomes
$$\tan. \text{TPF} = \frac{p}{y''} = \text{PTF}.$$
Hence the truth of the proposition is manifest, and FT=FP.

PROPOSITION V.

(52.) *If from the focus of a parabola a line be drawn perpendicular to the tangent, the point of intersection will be on the axis of ordinates.*

The equation of a line passing through a point is of the form
$$y - y'' = a'(x - x''). \quad (1)$$
If the line pass through the focus F, whose co-ordinates are $x'' = \tfrac{1}{2}p$, $y'' = 0$, the above equation becomes
$$y = a'(x - \tfrac{1}{2}p). \quad (2)$$
And the condition that this line shall be perpendicular to the tan-

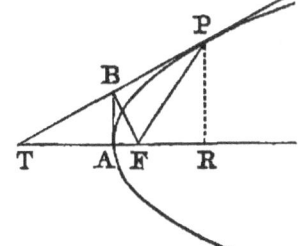

gent requires that the equation $a'a+1=0$ shall be fulfilled.

$$\therefore a' = -\frac{1}{a} = -\frac{y''}{p}.$$

Equation 2 becomes

$$y = -\frac{y''}{p}(x - \tfrac{1}{2}p). \qquad (3)$$

But the equation of the tangent line is

$$yy'' = p(x + x''). \qquad (4)$$

Combining equations (3) and (4), substituting for y''^2 its value $2px''$, and reducing, we have

$$x(2x'' + p) = 0;$$

whence $x = 0$, and $x'' = -\tfrac{1}{2}p$. Alg., Art. 105.

Therefore the point B, at which the perpendicular intersects the tangent, is on the axis of Y.

(53.) *Cor.* By Geom., B. IV., Theor. IX.,

FA : FB :: FB : FT;

or FA : FB :: FB : FP,

because FT = FP (Prop. IV.).

Hence *the perpendicular FB is a mean proportional between the distance from the focus to the vertex and the distance from the focus to the point of tangency.*

PROPOSITION VI.

(54.) *The equation of the parabola referred to a tangent line and the diameter passing through the point of tangency, the origin being the point of tangency, is*

$$y^2 = 2p'x,$$

in which $2p'$ is the parameter of the diameter passing through the origin.

The equation of the parabola referred to rectangular axes, of which A is the origin, is

$$y^2 = 2px. \qquad (1)$$

And the formulas for passing from these to a system of oblique axes, of which A' is the origin, are

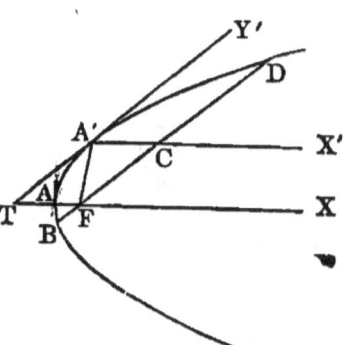

$x = a + x_{,} \cos.\ a + y_{,} \cos.\ a',$
$y = b + x_{,} \sin.\ a + y_{,} \sin.\ a',$
where a and b are the co-ordinates of the point A'.

Substitute these in (Eq. 1), we have
$b^2 + 2b(x_{,} \sin. a + y_{,} \sin. a') + x_{,}^2 \sin.^2 a + 2x_{,}y_{,} \sin. a \sin. a'$
$+ y_{,}^2 \sin.^2 a' = 2pa + 2px_{,} \cos. a + 2py_{,} \cos. a'$.   (2)

But, because the point A' is on the curve, the co-ordinates of that point will satisfy the equation of the curve, and we shall have
$$b^2 = 2pa.$$
And since the new axis $A'X'$ is parallel to the primitive axis AX, we shall have
$$a = 0.$$
$\therefore \sin. a = 0$, and $\cos. a = 1$.

Making these substitutions in (2) gives
$2by_{,} \sin. a' + y_{,}^2 \sin.^2 a' = 2px_{,} + 2py_{,} \cos. a'$.   (3)

Now the tangent of the angle which the tangent makes with the axis is $\dfrac{p}{y'}$, where y' is the ordinate of the point of tangency, and is therefore equal to b.

$$\therefore \frac{p}{y'} = \frac{p}{b} = \tan. a' = \frac{\sin. a'}{\cos. a'}.$$

$\therefore p \cos. a' = b \sin. a'$.

Substituting this value of $p \cos. a'$ in (3), we have
$2by' \sin. a' + y'^2 \sin.^2 a' = 2px' + 2by' \sin. a'$.

$$\therefore y'^2 = \frac{2px'}{\sin.^2 a'}, \text{ or } y_{,}^2 = \frac{2px_{,}}{\sin.^2 a'}.$$

Or, to simplify its form, put $\dfrac{2p}{\sin.^2 a'} = 2p'$, and dropping the accents on the variables,
$$y^2 = 2p'x.$$

In this equation, for every value we assign to x we shall have two values of y, numerically equal with contrary signs. The curve is therefore symmetrical with respect to the axis $A'X'$.

(55.) *This axis bisects all chords of the parabola which are parallel to the tangent line $A'Y'$, or axis of $A'Y'$.* It may therefore be considered as a diameter, and hence *all diameters of the parabola are parallel to each other, and the centre of the parabola is at an infinite distance from the vertex.*

Proposition VII.

(56.) *The parameter of any diameter is equal to four times the distance from the vertex of that diameter to the directrix, or four times the distance from the vertex to the focus.*

From the last Proposition we have
$$\frac{p}{b} = \frac{\sin. \, a'}{\cos. \, a'},$$
or
$$\frac{p^2}{b^2} = \frac{\sin.^2 a'}{1 - \sin.^2 a'}.$$
Clearing of fractions and reducing,
$$\sin.^2 a' = \frac{p^2}{b^2 + p^2};$$
but $b^2 = 2ap$, from the equation of the curve.
$$\therefore \sin.^2 a' = \frac{p}{2a + p}.$$
And from the last Proposition,
$$2p' = \frac{2p}{\sin.^2 a'}. \quad \therefore \frac{p}{\sin.^2 a'} = p'.$$
$$\therefore \sin.^2 a' = \frac{p}{p'}.$$
Equating these two values of $\sin.^2 a'$, we have
$$\frac{p}{2a + p} = \frac{p}{p'}.$$
$$\therefore p' = 2a + p,$$
and
$$2p' = 4(a + \tfrac{1}{2}p).$$

But $(a + \tfrac{1}{2}p) =$ to the distance from the vertex A' to the directrix, or from the vertex to the focus.

Hence the truth of the proposition is manifest.

Proposition VIII.

(57.) *The polar equation of the parabola, the pole being at the focus, is*
$$r = \frac{p}{1 \pm \cos. \, \theta},$$
where p represents half the parameter, and θ is the angle which the radius vector makes with the axis.

ON THE PARABOLA. 43

1. Let DB be the directrix, P any point in the curve. Draw the radius vector FP. Now, by the definition, FP=DP, and DP= BR; hence FP=BR.

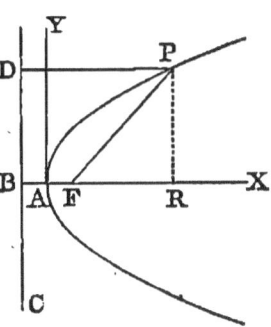

Put the radius vector, FP=r, and the angle PFR=θ. Then (Trig., Art. 5),
    FR=FP cos. θ=r cos. θ.
But   BR=BF+FR,
or r=p+r. cos. θ.

$$\therefore r = \frac{p}{1-\cos.\theta}.$$

If θ=0, cos. θ=1, and $r=\frac{p}{0}=\infty$,

the radius vector is equal to infinity. Now, as the angle θ increases from zero, the radius vector marks points in the curve. When θ=90°, cos. θ=0, and r=p. At this point the radius vector becomes the ordinate at the focus, and is equal to half the parameter. If θ=180°, cos. θ=-1, and $r=\frac{1}{2}p$, which marks the vertex A. In a manner entirely similar, the radius vector will describe the points of the curve below the axis; for when θ=270°, cos. θ=0, and r=p; when θ=360°, cos. θ=1, and $r=\frac{p}{0}=\infty$.

2. If we put the angle PFA=θ, then FR=$-r$. cos. θ, and r=$p-r$. cos. θ.

$$\therefore r = \frac{p}{1+\cos.\theta}.$$

In this case the values of r begin at the vertex A, and increase from the left to the right, from zero to 360.

PROPOSITION IX.

(58.) *The area of any portion of a parabola is equal to two thirds of the rectangle described on its abscissa and ordinate.*

Let ADB be a segment of a parabola referred to rectangular axes.

Bisect DB in E, and draw EQ parallel to DA; join AB. Through Q draw QN perpendicular to AD; then QN will be an ordinate to the diameter at the point N.

Now, because NDEQ is a rectangle, QN is equal to $ED = \tfrac{1}{2}DB$;

$\therefore \overline{QN}^2 : \overline{BD}^2 :: 1 : 4$.

But (Art. 40)
$AN : AD :: \overline{QN}^2 : \overline{BD}^2$;
$\therefore AN : AD :: 1 : 4$,
or $AN = \tfrac{1}{4}AD$;
hence ND or $QE = \tfrac{3}{4}AD$.

Again, since EG is parallel to DA, and $BE = \tfrac{1}{2}BD$,
$\therefore EG = \tfrac{1}{2}AD$; $\therefore QG = \tfrac{1}{4}AD$,
or $EG : GQ :: 2 : 1$.

Join AE, AQ, and QB; then, because BD is bisected in E, the triangle ABE is equal to half the triangle ABD; and since $GQ = \tfrac{1}{2}GE$, the triangle AGQ is equal to half the triangle AGE, and the triangle BGQ is equal to half the triangle BGE; hence the triangle $ABQ = \tfrac{1}{2}$ the triangle ABE, and, consequently, the triangle $ABQ = \tfrac{1}{4}$ the triangle ABD.

Now, suppose BE and DE were each bisected, and from the points of bisection lines were drawn parallel to DA, these lines would evidently bisect BG and GA respectively. Then, by drawing lines from the points in which those parallel lines would cut the curve to the extremities of the chords BQ and QA, the sum of the triangles formed within the spaces BQ and QA will be equal to $\tfrac{1}{4}$ the triangle ABQ, or $\tfrac{1}{16}$th of the triangle ABD.

By bisecting the halves of BE and ED, and drawing lines as before parallel to AD, and joining the points of their intersections with the curve to the extremities of the chords, a series of triangles would be formed in the remaining spaces, the sum of which would be equal to $\tfrac{1}{4}$th of the triangle immediately preceding it, or equal to $\tfrac{1}{64}$th of the triangle ABD.

Continue this bisection indefinitely, and we shall have the triangles filling the whole parabola, in which case the area of the parabola would ultimately be equal to the sum of the areas of all the triangles thus formed within it.

Put $a=$ the area of the $\triangle$ ABD.
Then the area of the parabola is $=a+\frac{1}{4}a+\frac{1}{16}a+\frac{1}{64}a+$, etc., *ad infin.*, which is a geometrical progression whose first term is a, and ratio $=\frac{1}{4}$, the sum of which (Alg., Art. 132) is

$$=\frac{a}{1-r}=\frac{a}{1-\frac{1}{4}}=\frac{4}{3}a.$$

Hence the area of the parabola is equal to *four thirds the area of the triangle* ABD. But the area of the triangle $ABD=\frac{1}{2}AD \times BD$.

$\therefore$ area par.$=\frac{2}{3}AD \times BD.$ Q. E. D.

ON THE ELLIPSE AND HYPERBOLA.

(59.)   *Definitions.*

1. An *Ellipse* is a plane curve, such that, if from any point in the curve two straight lines be drawn to two fixed points, the sum of these lines will be constantly equal to a given line.

2. The two fixed points are called the *foci*. If F' and F be two given fixed points, and A'A the given line, then, if F'P+FP be constantly equal to A'A for every position of the point P, the curve A'BAB' will be an ellipse.

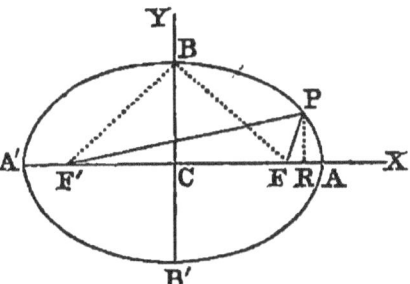

3. The distance of the point P from either focus is called the *radius vector*.

4. The point C at the middle of the straight line joining the foci is called the *centre* of the ellipse.

5. Any straight line drawn through the centre and limited by the curve is a *diameter*.

6. The points in which any diameter meets the curve are called the *vertices* of that diameter.

7. The diameter which passes through the foci is the

transverse diameter, and the diameter perpendicular to this is the *conjugate diameter*.

These are also called the *major* and *minor* axes.

8. A straight line which touches the curve in any point, but which does not cut it, is called a *tangent* to the curve at that point.

9. A diameter drawn parallel to the tangent at the vertex of any diameter is called the *conjugate diameter* to the latter, and the two diameters are called a *pair of conjugate diameters*.

10. If, at the extremity of the parameter of the transverse diameter, a tangent be drawn to meet the axis of X, and at the point of intersection a line be drawn perpendicular to that axis, the tangent is called the *focal tangent*, and the perpendicular line is called the *directrix*.

11. A line drawn perpendicular to the tangent at the point of tangency, and limited by the transverse axis, is called a *normal*.

12. The *subtangent* is the projection of the tangent on the axis of X, and

13. The *subnormal* is the projection of the normal on the same axis.

Proposition I.

(60.) *The equation of the ellipse referred to its centre and axes is*

$$A^2y^2 + B^2x^2 = A^2B^2,$$

where A and B are the semiaxes and x and y the general co-ordinates.

Let F and F' be the two fixed points, or the foci. Join FF', and bisect FF' in C. Let P be any point in the curve, and join FP, F'P. Draw PR and CY perpendicular to CX. Let C be the origin, and CX, CY the axes of co-ordinates.

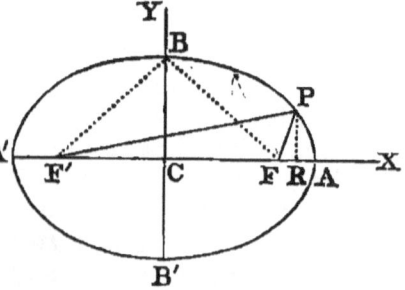

Let the line to which FP+F'P is constantly equal be denoted by 2A, and the distance FF' by 2c.

Then put $CR=x$, $RP=y$, $F'P=r'$, $FP=r$; then
$$r'^2=y^2+(c+x)^2, \quad (1)$$
$$r^2=y^2+(c-x)^2, \quad (2)$$
$$r'+r=2A. \quad (3)$$
Subtracting (2) from (1), we have
$$r'^2-r^2=4cx,$$
or $\qquad (r'+r)(r'-r)=4cx.$
But, because $\qquad r'+r=2A,$
$$\therefore r'-r=\frac{2cx}{A},$$
and $\qquad r'+r=2A.$
Combining these, we have
$$r'=A+\frac{cx}{A}; \quad \therefore r'^2=A^2+2cx+\frac{c^2x^2}{A^2},$$
$$r=A-\frac{cx}{A}; \quad \therefore r^2=A^2-2cx+\frac{c^2x^2}{A^2};$$
$$\therefore r'^2+r^2=2A^2+\frac{2c^2x^2}{A^2}.$$
Now, by adding (1) and (2) together, we have
$$r'^2+r^2=2(y^2+x^2+c^2).$$
Equate these two values of r'^2+r^2, and we obtain
$$y^2+x^2+c^2=A^2+\frac{c^2x^2}{A^2},$$
or $\qquad A^2y^2+(A^2-c^2)x^2=A^2(A^2-c^2). \quad (4)$

Since $r'+r$ or $2A$ is always greater than FF' or $2c$, therefore A is always greater than c, and hence the quantity A^2-c^2 is essentially positive.

Put $\qquad A^2-c^2=B^2;$
then $\qquad A^2y^2+B^2x^2=A^2B^2 \quad (5)$
is the equation of the ellipse referred to its centre and axes,
or $\qquad y^2=\frac{B^2}{A^2}(A^2-x^2). \quad (6)$
Solving the equations for y and x in succession, we have
$$y=\pm\frac{B}{A}\sqrt{A^2-x^2}, \quad (7)$$
$$x=\pm\frac{A}{B}\sqrt{B^2-y^2}. \quad (8)$$
When $x=0$, $y=\pm B$; and when $y=0$, $x=\pm A$.

Hence it appears that the curve cuts the axis of X at the points A to the right and A' to the left of the origin, at a distance from it equal to A, and that it cuts the axis of Y at a distance above and below the origin equal to B.

Hence, also, the quantities A and B are the semiaxes.

From equations (7) and (8) it is evident that the curve is symmetrical in regard to both axes. For we perceive that every value assumed for x gives two values for y numerically equal with opposite signs; and for every value we may assign to y, we shall have two values for x numerically equal with opposite signs.

(61.) And as BF is equal to BF', each being equal to A, it follows *that the distance from the extremity of the conjugate axis to either focus is equal to half the transverse axis.*

(62.) In Eq. 6, $\quad y^2 = \dfrac{B^2}{A^2}(A^2 - x^2).$

Let $\quad x = c = CF;$

then $\quad y^2 = \dfrac{B^2}{A^2}(A^2 - c^2).$

Now $\quad A^2 - c^2 = B^2,$ by Eq. 5;

$\therefore y^2 = \dfrac{B^4}{A^2};$

$\therefore A^2 y^2 = B^4,$ or $Ay = B^2.$

Hence B is a mean proportional between A and y. But y represents the ordinate at the focus. Putting the above into a proportion, we have

$$A : B :: B : y,$$
or $\quad 2A : 2B :: 2B : 2y.$

(63.) That is, *the double ordinate through the focus is a third proportional to the transverse and conjugate axes.* But the double ordinate through the focus is called the parameter; hence *the parameter of the transverse axis is* $= \dfrac{2B^2}{A}.$

(64.) The ratio of c to A is usually denoted by the symbol e, and is called the eccentricity of the ellipse. Thus, $\dfrac{c}{A} = e;$ $\therefore c = Ae,$ or $c^2 = A^2 e^2,$ and $2c = 2Ae.$ But

ON THE ELLIPSE AND HYPERBOLA. 49

$c^2 = A^2 - B^2$; $\therefore A^2 e^2 = A^2 - B^2$; $\therefore B^2 = A^2(1-e^2)$, or
$\dfrac{B^2}{A^2} = 1 - e^2$.

Substituting this in Eq. 6, we have
$$y^2 = (1-e^2)(A^2 - x^2).$$
We have also from the preceding
$$r' = A + \frac{cx}{A},$$
$$r = A - \frac{cx}{A}.$$

(65.) Let us substitute e for $\dfrac{c}{A}$, and then
$$r' = A + ex,$$
$$r = A - ex.$$
These last equations represent the distances from any point of the ellipse to either focus.

THE HYPERBOLA.

(66.) *An hyperbola* is a plane curve, such that, if from any point in the curve two straight lines be drawn to two given fixed points, the difference of these two lines will constantly be equal to a given line. The two given fixed points are called the *foci*.

Thus, if F′ and F are the two fixed points, and P any point in the hyperbola, then will
$$F'P - FP = A'A,$$
A′A being the given line, called the *transverse axis*. The middle point, C, of the straight line F′F, is called the centre of the hyperbola.

Any straight line drawn through the centre, and terminated by two opposite hyperbolas, is called a *diameter*, and the points in which it meets the hyperbolas are called the *vertices* of that diameter. The transverse diameter, when produced, passes through the foci, and the points in which it meets the curve are called the *principal vertices of the hyperbola*. The *parameter* of the

C

transverse axis is the double ordinate which passes through one of the foci.

Proposition II.

(67.) *The equation of the hyperbola referred to its centre and axes is*
$$A^2y^2 - B^2x^2 = -A^2B^2,$$
where A and B are the semi-axes, and x and y the general co-ordinates.

Let F' and F be the foci, and bisect $F'F$ in C, at which point draw the axis of Y perpendicular to $F'X$. Let P be any point of the curve, and draw PR perpendicular to the axis of X. Put 2A equal to the difference of the distances of the point P from the foci; and $PF' = r'$, $PF = r$, and $CF' = CF = c$. Also, let x and y denote the co-ordinates of the point P.

Then $F'P^2 = PR^2 + F'R^2$,
or
$$r'^2 = y^2 + (x+c)^2. \tag{1}$$
Again, $FP^2 = PR^2 + FR^2$,
or
$$r^2 = y^2 + (x-c)^2. \tag{2}$$
Adding (1) and (2), we have
$$r'^2 + r^2 = 2(y^2 + x^2 + c^2). \tag{3}$$
Subtracting (2) from (1), we have
$$r'^2 - r^2 = 4cx,$$
or
$$(r'+r)(r'-r) = 4cx. \tag{4}$$
But $r' - r = 2A$,
$$\therefore r' + r = \frac{2cx}{A}.$$

By addition and subtraction, the last two equations give us
$$r' = A + \frac{cx}{A}, \tag{5}$$

$$r = A - \frac{cx}{A}. \tag{6}$$

Squaring these values of r' and r, and substituting in (3), we obtain
$$A^2 + \frac{c^2x^2}{A^2} = y^2 + x^2 + c^2;$$
or, by reduction,
$$A^2y^2 + (A^2 - c^2)x^2 = A^2(A^2 - c^2). \tag{7}$$

ON THE ELLIPSE AND HYPERBOLA. 51

Now 2A must always be less than 2c, and therefore $A < c$.
Hence $A^2 - c^2$ is essentially negative. Assuming
$$A^2 - c^2 = -B^2,$$
Equation (7) becomes
$$A^2 y^2 - B^2 x^2 = -A^2 B^2,$$
or
$$y^2 = \frac{B^2}{A^2}(x^2 - A^2), \qquad (8)$$
which is the equation of the hyperbola referred to its centre and axes.

Solving the equation for x and y in succession, we obtain
$$x = \frac{A}{B}\sqrt{y^2 + B^2}, \qquad (9)$$
$$y = \frac{B}{A}\sqrt{x^2 - A^2}. \qquad (10)$$

Let $y = 0$; then $x = \pm A = CA$ or CA'.

Let $x = 0$; then $y = \pm B\sqrt{-1}$, an *imaginary expression*, from which it appears that the curve does not meet the axis of y. We may, however, take two points, B and B', on different sides of C, making
$$CB = CB' = \sqrt{c^2 - A^2}.$$

In order to determine the position of these points, with A as a centre and radius equal to CF'' or CF, describe a circle cutting the axis of y in two points, B and B'; these will be the points required. For
$$CB = \sqrt{CF'^2 - CA^2} = \sqrt{c^2 - A^2}.$$

The line BB', which is perpendicular to the transverse axis at its middle point, is called the *conjugate* axis.

(68.) If $B = A$, the equation of the hyperbola becomes
$$y^2 - x^2 = -A^2,$$
which is called the *equilateral hyperbola*.

From equations (9) and (10) we perceive that if x be less than A the values of y are imaginary; we therefore conclude that there is no point in the curve situate between A and A'. Again, if x be taken *negative* and equal to A, the value of y will be *zero*, and for every value of x negative and greater than A, y will have two values numerically equal with opposite signs. Hence we conclude that the hyperbola consists of two branches, one extending indefinitely to the right of A, and the other indefinitely to the left of A', and both symmetrically situated with regard to the axis of x.

The ratio of c to A is usually, as in the ellipse, denoted by the symbol e, and is called the *eccentricity* of the hyperbola. Thus,

$$\frac{c}{A}=e\,;\ \therefore c=Ae,\ \text{or}\ c^2=A^2 e^2.$$

But $\qquad c^2-A^2=A^2(e^2-1),$
where $\qquad c^2=A^2+B^2,$
or $\qquad B=\pm A(e^2-1)^{\frac{1}{2}}.$

(69.) In order to transfer the origin from the centre to the vertex of the transverse axis, we have simply to substitute $(x+A)$ for A in (8), and reduce. We then have

$$y^2=\frac{B^2}{A^2}(2Ax+x^2),$$

the equation of the hyperbola referred to the extremity of the transverse axis as the origin of co-ordinates.

(70.) If in (8) we make $x=c=CF$, then

$$y^2=\frac{B^2}{A^2}(c^2-A^2)=\frac{B^2}{A^2}\times B^2\,;$$

$$\therefore y=\frac{B^2}{A}.$$

Putting this into a proportion, we have
$\qquad\qquad A:B::B:y,$
or $\qquad\qquad 2A:2B::2B:2y.$

But $2y$ is the double ordinate passing through a focus, and is called the parameter of the transverse axis; therefore,

(71.) *The parameter of the transverse axis is a third proportional to that axis and its conjugate.*

(72.) In the equations $r'=A+\dfrac{cx}{A}$ and $r=-A+\dfrac{cx}{A}$, substitute e for $\dfrac{c}{A}$, and we have $r'=A+ex$ and $r=-A+ex$ for the distance from the focus to any point on the hyperbola.

(73.) The properties of the hyperbola may be divided into two classes: in the first class may be comprehended all such properties as are analogous to those of the ellipse, and in the second class all such as are derived from its relation to the asymptotes. We shall therefore discuss the analogous properties of the ellipse and hyperbola together, but the properties of the hyperbola which are derived from its relation to the asymptotes we shall discuss separately.

PROPOSITION III.

(74.) *In an ellipse or hyperbola the squares of the ordinates are to each other as the rectangles of the segments from the foot of each ordinate respectively to the vertices of the transverse axis.*

The equation of the ellipse referred to its centre and axis is
$$y^2 = \frac{B^2}{A^2}(A^2 - x^2);\qquad(1)$$
and that of the hyperbola is
$$y^2 = \frac{B^2}{A^2}(x^2 - A^2).\qquad(2)$$

Now, if we designate a particular ordinate by y' and its corresponding abscissa by x', we shall have

for the ellipse $\quad y'^2 = \dfrac{B^2}{A^2}(A^2 - x'^2);\qquad(3)$

for the hyper. $\quad y'^2 = \dfrac{B^2}{A^2}(x'^2 - A^2).\qquad(4)$

And if we designate a second ordinate by y'' and its corresponding abscissa by x'', we shall have

for the ellipse $\quad y''^2 = \dfrac{B^2}{A^2}(A^2 - x''^2);\qquad(5)$

for the hyper. $\quad y''^2 = \dfrac{B^2}{A^2}(x''^2 - A^2).\qquad(6)$

Dividing (3) by (5), member by member, we obtain
$$\frac{y'^2}{y''^2} = \frac{A^2 - x'^2}{A^2 - x''^2} = \frac{(A+x')(A-x')}{(A+x'')(A-x'')};\qquad(7)$$
$$\therefore y'^2 : y''^2 :: (A+x')(A-x') : (A+x'')(A-x''),\qquad(8)$$
where $(A+x')$ and $(A-x')$ are the segments corresponding to y', and $(A+x'')$ and $(A-x'')$ are the segments corresponding to y'' in the ellipse.

Again, dividing (4) by (6), we shall have for the hyperbola

$$\frac{y'^2}{y''^2}=\frac{x'^2-A^2}{x''^2-A^2}=\frac{(x'+A)(x'-A)}{(x''+A)(x''-A)};\qquad (9)$$

$$\therefore y'^2:y''^2::(x'+A)(x'-A):(x''+A)(x''-A),\qquad (10)$$

where $(x'+A)$ and $(x'-A)$ are the segments corresponding to y', and $(x''+A)$, $(x''-A)$ are the segments corresponding to y'' in the hyperbola; whence the truth of the proposition is manifest.

Proposition IV.

(75.) *If through the vertices of the transverse axis two supplementary chords be drawn, the product of the tangents of the angles which they form with it on the same side will be numerically equal to the square of the ratio of the semi-axes; in the ellipse it is negative, and in the hyperbola positive.*

The equation of a line passing through a point whose co-ordinates are $x'=A$, $y'=0$, is

$$y=a(x-A).$$

The equation of another line passing through a point whose co-ordinates are $x''=-A$, $y''=0$, is

$$y=a'(x+A).$$

As these lines are required to intersect on a curve, their equations must not only exist simultaneously, but also with the equation of the curve on which they intersect. Multiplying the two equations together, we have

$$y^2=aa'(x^2-A^2).\qquad (a)$$

But the equation of the ellipse is

$$y^2=\frac{B^2}{A^2}(A^2-x^2),$$

or

$$=-\frac{B^2}{A^2}(x^2-A^2).\qquad (b)$$

Equating these two values of y^2, we have

$$aa'=-\frac{B^2}{A^2}.$$

Again, the equation of the hyperbola is

$$y^2=\frac{B^2}{A^2}(x^2-A^2).\qquad (c)$$

Equating this value of y^2 with that in (a), we obtain

$$aa' = \frac{B^2}{A^2},$$

whence the truth of the proposition is manifest.

PROPOSITION V.

(76.) *To find the equation of a tangent line to an ellipse or hyperbola.*

Draw any line, P' P", cutting the curve in the points P', P"; if this line be moved toward P it will become a tangent when the points coalesce. Let the co-ordinates of the point P' be denoted by x', y', and those of the point P" by x'', y''. 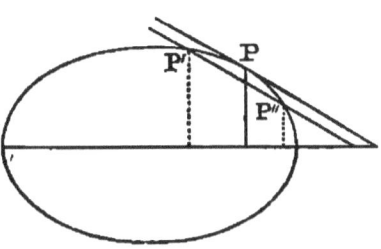 The equation of a line passing through these two points will be

$$y - y'' = \frac{y'' - y'}{x'' - x'}(x - x''). \tag{1}$$

Since the points are on the curve, we shall have for the equations of the curves

$$A^2 y'^2 \pm B^2 x'^2 = \pm A^2 B^2; \tag{2}$$
$$A^2 y''^2 \pm B^2 x''^2 = \pm A^2 B^2. \tag{3}$$

Subtracting (2) from (3), we have
$$A^2(y''^2 - y'^2) \pm B^2(x''^2 - x'^2) = 0,$$
or $\quad A^2(y'' + y')(y'' - y') \pm B^2(x'' + x')(x'' - x') = 0.$
$$\therefore \frac{y'' - y'}{x'' - x'} = \mp \frac{B^2(x'' + x')}{A^2(y'' + y')}.$$

Substituting this value in (1), the equation of the secant line becomes

$$y - y'' = \mp \frac{B^2(x'' + x')}{A^2(y'' + y')}(x - x'').$$

But when the secant becomes a tangent the points coalesce, and then $x' = x''$ and $y' = y''$; this reduces the last equation to

$$y - y'' = \mp \frac{B^2 x''}{A^2 y''}(x - x'').$$

The term $\dfrac{B^2 x''}{A^2 y''}$ is the tangent of the angle which the tangent line makes with the axis of x; the upper sign $(-)$ belonging to the ellipse, the lower sign $(+)$ belonging to the hyperbola.

Clearing of fractions, we have
$$A^2yy'' - A^2y''^2 = -B^2xx'' + B^2x''^2,$$
or $\quad A^2yy'' + B^2xx'' = A^2y''^2 + B^2x''^2 = A^2B^2;$
$\therefore A^2yy'' + B^2xx'' = A^2B^2,$

the equation of the tangent to the ellipse.

Using the other sign, we have
$$A^2yy'' - A^2y''^2 = B^2xx'' - B^2x''^2,$$
or $\quad A^2yy'' - B^2xx'' = A^2y''^2 - B^2x''^2 = -A^2B^2;$
$\therefore A^2yy'' - B^2xx'' = -A^2B^2,$

the equation of a tangent line to the hyperbola.

(77.) To find the point in which the tangent intersects the axis of x, make $y=0$ in the equation of the tangent line; then for the ellipse we have

$$x = \frac{A^2}{x''}$$

which is equal to the distance from the centre of the ellipse to the point required; and to find the length of the subtangent, we must subtract the abscissa of the point of tangency from this. That is,

$$subtangent = \frac{A^2}{x''} - x'' = \frac{A^2 - x''^2}{x''}.$$

(78.) As this expression for the subtangent is independent of the conjugate axis, the subtangent is therefore the same for all ellipses having the same transverse diameter, and consequently belongs to the circle described upon the transverse axis.

From this we are enabled to draw a tangent to an ellipse at any given point. Let P be the given point, and on the transverse axis describe the circle A'P'A. Through P draw the ordinate PR, and produce it to meet the circle in P'. From P' draw the tangent to

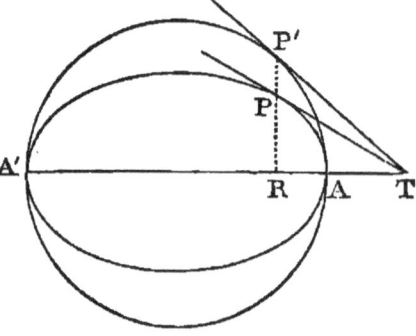

the circle, cutting the axis of x in T; join PT, which will be the tangent required.

(79.) To find the subtangent of the hyperbola: as the expression $\dfrac{A^2}{x''}$ is the distance from the centre of the curve to the point in which the tangent cuts the axis of x, we must subtract this from the abscissa of the point of tangency. That is,

$$\textit{subtangent}=x''-\dfrac{A^2}{x''}=\dfrac{x''^2-A^2}{x''}.$$

(80.) Since the normal is perpendicular to the tangent at the point of tangency, the tangent of the angle which it forms with the axis of x must be the reciprocal of that which the tangent line makes with the same axis, and also have a contrary sign.

But the tangent of the angle which the tangent line makes with the axis of x is $-\dfrac{B^2 x''}{A^2 y''}$; therefore the tangent of the angle which the normal makes with the axis of x in the ellipse is

$$\dfrac{A^2 y''}{B^2 x''}.$$

Substituting this in (1), we have

$$y-y''=\dfrac{A^2 y''}{B^2 x''}(x-x'')$$

for the *equation of the normal line to the ellipse*.

(81.) To find the *subnormal*, we must first find the point in which the normal cuts the axis of x. This is done by making $y=0$; then, by a little reduction, we have

$$CN = x = \dfrac{A^2-B^2}{A^2} x''.$$

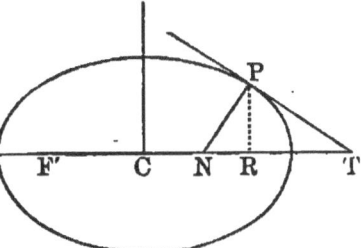

Now, subtracting this from CR, the abscissa of the point of tangency, which is represented by x'', we shall have

$$NR = x'' - \dfrac{A^2-B^2}{A^2} x'',$$

or
$$NR = \frac{B^2 x''}{A^2} = \text{subnormal}.$$

(82.) If we put $\frac{A^2 - B^2}{A^2} = e^2$, we shall have
$$CN = e^2 x''.$$
And adding c to this, which is $= Ae$, we get
$$F'N = Ae + e^2 x''$$
$$= e(A + ex'')$$
for *the distance from the focus F' to the foot of the normal.*

(83.) In a manner entirely similar, by taking, with a contrary sign, the reciprocal of the tangent of the angle which a tangent line to the hyperbola makes with the axis of x, we shall have
$$y - y'' = -\frac{A^2 y''}{B^2 x''}(x - x'')$$
as *the equation of a normal to the hyperbola.*

(84.) Making $y = 0$, we shall have, after a little reduction,
$$CN = x = \frac{A^2 + B^2}{A^2} x''.$$

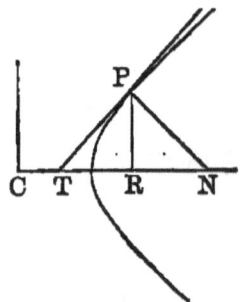

And subtracting CR, which is x'', we will have
$$RN = \frac{B^2 x''}{A^2} = \text{subnormal}.$$
Also, if we put $\frac{A^2 + B^2}{A^2} = e^2$, we shall have
$$CN = e^2 x''.$$
And by adding c, or its equal Ae, to this,
$$F'N = Ae + e^2 x''$$
$$= e(A + ex'')$$
for *the distance from the focus to the foot of the normal.*

Proposition VI.

(85.) *If two circles be described, one on the transverse, the other on the conjugate axis of an ellipse, then any ordinate of the circle drawn to the transverse axis will be to the corresponding ordinate of the ellipse as the semi-transverse axis is to the semi-conjugate axis; and any ordinate of the circle drawn to the conjugate axis will be to the corresponding ordinate of the ellipse as the semi-conjugate axis is to the semi-transverse.*

Let x' denote the abscissa of the points P and P', y' the ordinate of the point P, and Y' the ordinate of the point P'. Then the equation of the ellipse will be

$$y'^2 = \frac{B^2}{A^2}(A^2 - x'^2), \quad (1)$$

and the equation of the circle will be

$$Y'^2 = A^2 - x'^2. \quad (2)$$

Then, by dividing (2) by (1), member by member,

$$\frac{Y'^2}{y'^2} = \frac{A^2}{B^2};$$

$$\therefore \frac{Y'}{y'} = \frac{A}{B};$$

or $\quad Y' : y' :: A : B.$

(86.) Again: let $RP = x'$, and $RP' = X'$, y' the ordinate, we shall have

$$x'^2 = \frac{A^2}{B^2}(B^2 - y'^2),$$

eq. of the ellipse;
and $\quad X'^2 = B^2 - y'^2,$
eq. of the circle.

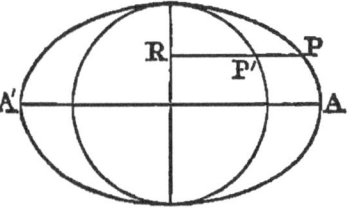

$$\therefore \frac{X'^2}{x'^2} = \frac{B^2}{A^2},$$

or $\quad \dfrac{X'}{x'} = \dfrac{B}{A};$

whence $\quad X' : x' :: B : A,$

from which the truth of the proposition is manifest.

PROPOSITION VII.

(87.) *If a line be drawn tangent to an ellipse or hyperbola, and two lines be drawn from the point of tangency to the foci, these lines will make equal angles with the tangent line.*

Let P be any point of the ellipse at which a tangent line, TT', is drawn, and PF', PF two lines drawn from the point of tangency to the foci; then will the angle FPT be equal to the angle F'PT'. Bisect the angle F'PF by the line PN. Then (Geom., B. IV., Theor. V.),

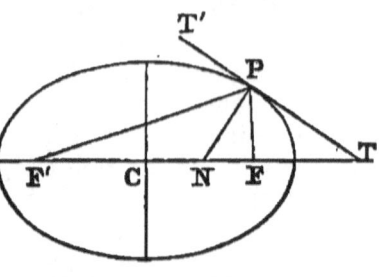

$$FN : NF' :: FP : F'P ;$$
or $\quad FN+NF' : NF' :: FP+F'P : F'P$, by composition.
But $\quad FN+NF'=2c=2Ae$, Prop. I., Art. (64),
$$FP+F'P=2A,$$
and $\quad F'P=A+ex''$, Prop. I., Art. (65).
$$\therefore 2Ae : NF' :: 2A : A+ex'' ;$$
$$\therefore NF'=e(A+ex'').$$

But by Prop. V., Art. (82), $e(A+ex'')$ denotes the distance from the focus F' to the foot of the normal; hence PN is a normal, and is perpendicular to the tangent; and because PN bisects the angle F'PF, therefore the angle F'PT' must be equal to the angle FPT, or the lines PF', PF make equal angles with the tangent.

(88.) In a manner entirely similar we may prove the same property in the hyperbola.

Produce F'P, and bisect the exterior angle FPM by the line PN. Then (Geom., B. IV., Theor. V., Case 2),
$$NF' : NF :: F'P : FP ; \quad (1)$$
and by division,
$$NF'-NF : NF' :: F'P-FP : F'P. \quad (2)$$
But $\quad NF'-NF=2c=2Ae$,
$$F'P-FP=2A,$$
and $\quad F'P=A+ex''$.

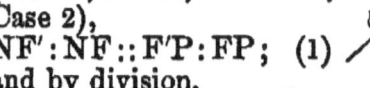

Substituting these in (2), we have
$$2Ae : NF' :: 2A : A+ex'',$$
whence $\quad NF'=e(A+ex'')$.

But by Prop. V., Art. (84), $e(A+ex'')$ denotes the distance from the focus F' to the foot of the normal; hence PN is a normal; and the equal angles FPN and NPM, each being taken from the right angles TPN or T'PN, leaves the angle TPF equal to the angle T'PM; but the angle TPM is equal to the angle F'PT; therefore the angle T'PF is equal to the angle TPF'.

(89.) The relation between the angles formed by the tangent line and lines drawn from the point of tangency to the foci enables us to draw a tangent to the hyperbola at any given point. Thus, let P be the given point; draw the two lines PF', PF, from the point of tangency to the foci. On PF' make PG equal to PF, and draw PT perpendicular to GF; PT will be the tangent required. The truth of this construction is evident.

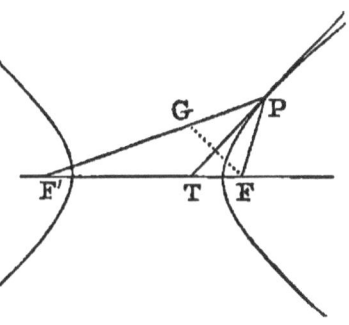

(90.) THE ELLIPSE AND HYPERBOLA REFERRED TO THEIR CENTRE AND CONJUGATE DIAMETERS.

Two diameters are said to be conjugate to each other in an ellipse or hyperbola when either of them is parallel to the two tangents which may be drawn through the vertices of the other.

Since two supplementary chords may always be drawn respectively parallel to a diameter and the tangent lines drawn through the vertices of that diameter, it follows that two supplementary chords may always be drawn respectively parallel to two conjugate diameters.

If we designate the tangents of the angles which two conjugate diameters make with the transverse axis by a and a', then we must have in the ellipse

$$aa' = -\frac{B^2}{A^2}, \tag{1}$$

and in the hyperbola

$$aa' = \frac{B^2}{A^2}. \tag{2}$$

Let us designate the angle whose tangent is a by α, and the angle whose tangent is a' by α'; then
$$a = \tan. \alpha = \frac{\sin. \alpha}{\cos. \alpha},$$
$$a' = \tan. \alpha' = \frac{\sin. \alpha'}{\cos. \alpha'}.$$
Substitute this in (1), and we have
$$\frac{\sin. \alpha}{\cos. \alpha} \cdot \frac{\sin. \alpha'}{\cos. \alpha'} = -\frac{B^2}{A^2}, \qquad (3)$$
or $\qquad A^2 \sin. \alpha \sin. \alpha' + B^2 \cos. \alpha \cos. \alpha' = 0, \qquad (4)$
for the equation of condition for conjugate diameters in an ellipse.

Again: substituting the above values of a and a' in (2), we have
$$\frac{\sin. \alpha}{\cos. \alpha} \cdot \frac{\sin. \alpha'}{\cos. \alpha'} = \frac{B^2}{A^2}, \qquad (5)$$
or $\qquad A^2 \sin. \alpha \sin. \alpha' - B^2 \cos. \alpha \cos. \alpha' = 0, \qquad (6)$
for the equation of condition for conjugate diameters in an hyperbola.

In (4) or (6), if we make $\alpha = 0$, we shall have
$$B^2 \cos. \alpha' = 0;$$
$$\therefore \alpha' = 90°.$$

(91.) That is, if one of the conjugate diameters, either in the ellipse or the hyperbola, coincides with the transverse axis, the other will coincide with the conjugate axis. Hence the axes fulfill the condition of conjugate diameters, as they should do, since each is parallel to the two tangents which may be drawn through the vertices of the other.

Let us ascertain whether there are other conjugate diameters at right angles to each other.

In order that they may be at right angles, the difference between the angles which each makes with the transverse axis must be equal to 90°, or
$$\alpha' - \alpha = 90°, \text{ or } \alpha' = 90° + \alpha;$$
$\sin. \alpha' = \sin.(90° + \alpha) = \sin. 90° \cos. \alpha + \sin. \alpha \cos. 90°$
$\qquad = +\cos. \alpha;$
$\cos. \alpha' = \cos.(90° + \alpha) = \cos. 90° \cos. \alpha - \sin. \alpha \sin. 90°$
$\qquad = -\sin. \alpha.$

If we substitute these values in (4) and (6), we shall have

ON THE ELLIPSE AND HYPERBOLA. 63

$$A^2 \sin. \alpha \cos. \alpha - B^2 \sin. \alpha \cos. \alpha = 0,$$
or
$$(A^2 - B^2)\sin. \alpha \cos. \alpha = 0,$$
which can not be satisfied unless
$$\alpha = 0 \text{ or } 90°,$$
and shows that the axes are the only conjugate diameters that are at right angles to each other.

PROPOSITION VII.

(92.) *To find the equation of the ellipse or hyperbola referred to its centre and conjugate diameters.*

In the equation
$$A^2 y^2 \pm B^2 x^2 = \pm A^2 B^2, \qquad (1)$$
the upper signs indicate the equation of the ellipse, the lower signs the equation of the hyperbola.

Take the formulas for passing from a system of rectangular to oblique co-ordinates, which are
$$x = x_{,} \cos. \alpha + y_{,} \cos. \alpha',$$
$$y = x_{,} \sin. \alpha + y_{,} \sin. \alpha'.$$

Square these values of x and y, and substitute them in (1), we shall have

$$\left\{\begin{array}{l}(A^2 \sin.^2 \alpha' \pm B^2 \cos.^2 \alpha')y_{,}^2 + \\ (A^2 \sin.^2 \alpha \pm B^2 \cos.^2 \alpha)x_{,}^2 + \\ 2(A^2 \sin. \alpha \sin. \alpha' \pm B^2 \cos. \alpha \cos. \alpha')x_{,}y_{,}\end{array}\right\} = \pm A^2 B^2. \quad (2)$$

But the equation of condition that the new axes shall be conjugate diameters gives
$$A^2 \sin. \alpha \sin. \alpha' \pm B^2 \cos. \alpha \cos. \alpha' = 0.$$
Hence the third term of equation (2) vanishes, and we have
$$\left\{\begin{array}{l}(A^2 \sin.^2 \alpha' \pm B^2 \cos.^2 \alpha')y_{,}^2 + \\ (A^2 \sin.^2 \alpha \pm B^2 \cos.^2 \alpha)x_{,}^2\end{array}\right\} = \pm A^2 B^2. \quad (3)$$

In order to find the points in which either curve cuts the axis of $x_{,}$ or the axis of $y_{,}$, we make $y_{,} = 0$ and $x_{,} = 0$ in succession. Therefore, if $y_{,} = 0$, we have
$$(A^2 \sin.^2 \alpha \pm B^2 \cos.^2 \alpha)x_{,}^2 = \pm A^2 B^2,$$
or
$$x_{,}^2 = \frac{\pm A^2 B^2}{A^2 \sin.^2 \alpha \pm B^2 \cos.^2 \alpha} = \overline{CD}^2.$$

And if $x_{,} = 0$, then
$$y_{,}^2 = \frac{\pm A^2 B^2}{A^2 \sin.^2 \alpha' \pm B^2 \cos.^2 \alpha'} = \overline{CE}^2.$$

Taking the upper sign for the ellipse, making $\overline{CD}^2 =$

$\overline{A'^2}$, $\overline{CE}^2 = B'^2$, and finding the value of the denominators,

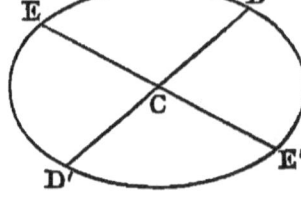

$$A^2 \sin^2 a + B^2 \cos^2 a = \frac{A^2 B^2}{A'^2},$$

$$A^2 \sin^2 a' + B^2 \cos^2 a' = \frac{A^2 B^2}{B'^2},$$

and substituting these in (3), we shall have

$$\frac{A^2 B^2}{B'^2} y_{,}^2 + \frac{A^2 B^2}{A'^2} x_{,}^2 = A^2 B^2;$$

which, by making the co-ordinates general, becomes

$$A'^2 y^2 + B'^2 x^2 = A'^2 B'^2, \qquad (4)$$

the equation of the ellipse referred to its centre and conjugate diameters.

(93.) Again: taking the lower sign for the hyperbola, we have

$$x_{,}^2 = \frac{-A^2 B^2}{A^2 \sin^2 a - B^2 \cos^2 a} = \overline{CD}^2,$$

$$y_{,}^2 = \frac{-A^2 B^2}{A^2 \sin^2 a' - B^2 \cos^2 a'} = \overline{CE}^2.$$

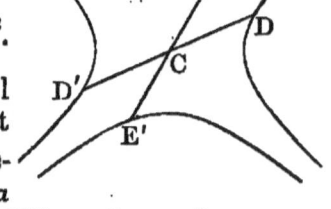

Now if $\overline{CD}^2$ is positive, $\overline{CE}^2$ will be negative; for, in order that $\overline{CD}^2$ shall be positive, the denominator $A^2 \sin^2 a - B^2 \cos^2 a$ must be negative, or $A^2 \sin^2 a < B^2 \cos^2 a$, and

$$\frac{\sin^2 a}{\cos^2 a} < \frac{B^2}{A^2}, \text{ or } \frac{\sin a}{\cos a} < \frac{B}{A}.$$

But $\quad \dfrac{\sin a}{\cos a} \cdot \dfrac{\sin a'}{\cos a'} = \dfrac{B^2}{A^2};$

hence $\quad \dfrac{\sin a'}{\cos a'} > \dfrac{B}{A},$

and $\quad \dfrac{\sin^2 a'}{\cos^2 a'} > \dfrac{B^2}{A^2},$

or $\quad A^2 \sin^2 a' > B^2 \cos^2 a'.$

The denominator $(A^2 \sin^2 a' - B^2 \cos^2 a')$ is therefore positive; and as the numerator is negative, $\therefore \overline{CE}^2$ must be negative. If we put $\overline{CD}^2 = A'^2$, and $\overline{CE}^2 = -B'^2$, we shall have from (3)

$$\frac{A^2B^2}{-B'^2}y_{\prime}^2 + \frac{A^2B^2}{A'^2}x_{\prime}^2 = +A^2B^2;$$

$$\therefore A'^2 y_{\prime}^2 - B'^2 x_{\prime}^2 = -A'^2 B'^2;$$

or, making the co-ordinates general,

$$A'^2 y^2 - B'^2 x^2 = -A'^2 B'^2,$$

which is *the equation of the hyperbola referred to its centre and conjugate diameters.*

(94.) We see that the equation of the ellipse, as well as the equation of the hyperbola, when referred to the centre and conjugate diameters, is of the same form as that when referred to the centre and axes; it follows, therefore, that every value of x will give two values of y numerically equal, but with opposite signs; or, if B' were real, every value of y would give two equal values of x with opposite signs; hence the curves are symmetrical with respect to the diameter which it intersects. That is, *either diameter will bisect all chords drawn parallel to the other and terminated by the curve.*

Proposition VIII.

(95.) *The squares of the ordinates to either of the conjugate diameters of an ellipse or hyperbola are to each other as the rectangles of the corresponding segments from the foot of the ordinates respectively to the vertices of the diameter.*

The equation of the curves is

$$A'^2 y^2 \pm B'^2 x^2 = \pm A'^2 B'^2,$$

where the upper signs belong to the ellipse, and the lower signs to the hyperbola.

If we designate any two ordinates by y' and y'', the corresponding abscissas by x', x'', we shall have for the ellipse

$$\frac{y'^2}{y''^2} = \frac{(A'+x')(A'-x')}{(A'+x'')(A'-x'')},$$

and for the hyperbola

$$\frac{y'^2}{y''^2} = \frac{(x'+A')(x'-A')}{(x''+A')(x''-A')}.$$

(96.) The *parameter* of any diameter is a third pro-

portional to the diameter and its conjugate. Thus, if P denote the parameter of the diameter 2A', we shall have
$$2A' : 2B' :: 2B' : P,$$
or
$$P = \frac{2B'^2}{A'}.$$

(97.) The parameter of the transverse axis is equal to
$$\frac{2B^2}{A},$$
and the parameter of the conjugate axis is equal to
$$\frac{2A^2}{B}.$$

PROPOSITION IX.

(98.) *To find the equation of a tangent line to the ellipse or hyperbola referred to the conjugate diameters.*

Let x', y' and x'', y'' be the co-ordinates of two points on the curve; then the equation of a right line passing through these points will be

$$y - y'' = \frac{y'' - y'}{x'' - x'}(x - x''). \qquad (1)$$

But, as the points are on the curve, we shall have
$$A'^2 y'^2 \pm B'^2 x'^2 = \pm A'^2 B'^2, \qquad (2)$$
$$A'^2 y''^2 \pm B'^2 x''^2 = \pm A'^2 B'^2. \qquad (3)$$

Subtracting (2) from (3), we obtain
$$A'^2(y'' + y')(y'' - y') \pm B'^2(x'' + x')(x'' - x') = 0; \qquad (4)$$
whence
$$\frac{y'' - y'}{x'' - x'} = \mp \frac{B'^2(x'' + x')}{A'^2(y'' + y')}.$$

Substituting this value in the equation of the secant line, it becomes

$$y - y'' = \mp \frac{B'^2(x'' + x')}{A'^2(y'' + y')}(x - x''). \qquad (5)$$

The secant will become a tangent when the two points coalesce, in which case $x'' = x'$, $y'' = y'$.

Making this substitution, (Eq. 5) becomes
$$y - y'' = \mp \frac{B'^2 x''}{A'^2 y''}(x - x'');$$

or, by reduction,
$$A'^2 y y'' \pm B'^2 x x'' = \pm A'^2 B'^2,$$
which is the most simple form of the equation of a tan-

gent line. The upper signs belong to the ellipse, the lower to the hyperbola.

PROPOSITION X.

(99.) In an ellipse, *the sum of the squares of any two conjugate diameters is equal to the sum of the squares of the axes.*

(100.) But in an hyperbola, *the difference of the squares of any two conjugate diameters is equal to the difference of the squares of the axes.*

Let $D'D$ and $E'E$ be any two conjugate diameters, x', y' the co-ordinates of D, x'', y'' the co-ordinates of E, a the angle DCA, aud a' the angle ACE; then

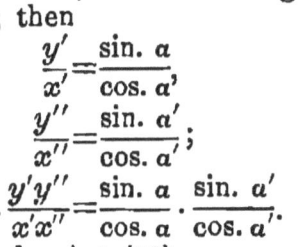

$$\frac{y'}{x'} = \frac{\sin. a}{\cos. a},$$
$$\frac{y''}{x''} = \frac{\sin. a'}{\cos. a'};$$
$$\therefore \frac{y'y''}{x'x''} = \frac{\sin. a}{\cos. a} \cdot \frac{\sin. a'}{\cos. a'}.$$

But, by Art. (90),
$$\frac{\sin. a}{\cos. a} \cdot \frac{\sin. a'}{\cos. a'} = \mp \frac{B^2}{A^2}.$$

The upper sign for the ellipse, the lower sign for the hyperbola.
$$\therefore \frac{y'y''}{x'x''} = \mp \frac{B^2}{A^2};$$

or, by squaring both members and reducing,
$$A^4 y'^2 y''^2 = B^4 x'^2 x''^2. \qquad (1)$$

The equation of the hyperbola is
$$A^2 y^2 - B^2 x^2 = -A^2 B^2, \qquad (a)$$

and the equation of the hyperbola conjugate to this is
$$B^2 x^2 - A^2 y^2 = -A^2 B^2, \qquad (b)$$

which is obtained by merely changing A into B and x into y in Eq. (a).

Now, since D and E are on conjugate hyperbolas, we shall have for the point D
$$A^2 y'^2 = \mp B^2 x'^2 \pm A^2 B^2; \qquad (2)$$

and for the point E
$$A^2y''^2 = \mp B^2x''^2 + A^2B^2. \quad (3)$$
Multiplying (2) and (3) together, we obtain
$$A^4y'^2y''^2 = B^4x'^2x''^2 - A^2B^4x''^2 \mp A^2B^4x'^2 \pm A^4B^4. \quad (4)$$
Equating these values of $A^4y'^2y''^2$, we have
$$B^4x'^2x''^2 = B^4x'^2x''^2 - A^2B^4x''^2 \mp A^2B^4x'^2 \pm A^4B^4; \quad (5)$$
or
$$0 = -x''^2 \mp x'^2 \pm A^2.$$
The upper sign belonging to the ellipse;
$$\therefore A^2 = x'^2 + x''^2, \quad (6)$$
and $\quad A^2 = x'^2 - x''^2$ for the hyperbola. $\quad (7)$

If, in Eqs. (2), (3), we arrange them thus,
$$\mp B^2x'^2 = A^2y'^2 \mp A^2B^2,$$
$$\mp B^2x''^2 = A^2y''^2 - A^2B^2,$$
we shall obtain
$$B^2 = y'^2 + y''^2 \text{ for the ellipse}, \quad (8)$$
and $\quad B^2 = y''^2 - y'^2$ for the hyperbola. $\quad (9)$

Adding (6) and (8) together, we have
$$A^2 + B^2 = x'^2 + y'^2 + x''^2 + y''^2 = A'^2 + B'^2. \quad (10)$$
Subtracting (9) from (7), we have
$$A^2 - B^2 = x'^2 + y'^2 - (x''^2 + y''^2) = A'^2 - B'^2. \quad (11)$$

(101.) Eq. (7) gives $\quad x'^2 = A^2 + x''^2,$
and (b) $\quad A^2y''^2 = B^2(A^2 + x''^2).;$

$$\therefore B^2x'^2 = A^2y''^2, \text{ or } x'^2 = \frac{A^2}{B^2}y''^2, \text{ or } x' = \frac{A}{B}y''.$$

In like manner, $\quad y' = \frac{B}{A}x''.$

Proposition XI.

(102.) *In an ellipse or hyperbola, the parallelogram which is formed by drawing tangent lines through the vertices of two conjugate diameters is equivalent to the rectangle formed by drawing tangent lines through the vertices of the axes.*

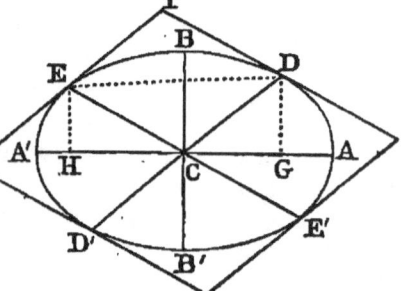

First. In the ellipse, let the parallelogram be drawn through the vertices D, D', E, E' of the two conjugate diameters DD' and EE'.

Join DE, and let fall the perpendiculars DG and EH; then will the area of the triangle DCE be equal to the area of the trapezoid GDEH minus the area of the two triangles CGD and CHE.

Put $CG = x'$, $GD = y'$; $CH = x''$, $HE = y''$;
then the area of the trapezoid $= \tfrac{1}{2}(x'' + x')(y'' + y')$,
the area of the triangle $CGD = \tfrac{1}{2}x'y'$,
the area of the triangle $CHE = \tfrac{1}{2}x''y''$;
$\therefore$ area of the triangle $DCE = \tfrac{1}{2}(x'y'' + x''y')$.

But the area of the triangle DCE is equal to one half of the parallelogram CDTE;
hence the area of $CDTE = x'y'' + x''y'$.

Now $\qquad y'' = \dfrac{B}{A}x'$,

and $\qquad x'' = \dfrac{A}{B}y'$, Art. (101);

$$\therefore CDTE = x'\dfrac{Bx'}{A} + y'\dfrac{Ay'}{B}$$

$$= \dfrac{A^2 y'^2 + B^2 x'^2}{AB} = \dfrac{A^2 B^2}{AB} = AB.$$

But the parallelogram which is formed by drawing tangent lines through the vertices of the two conjugate diameters is four times CDTE; therefore it is equal to $4AB = 2A \times 2B$, where 2A and 2B are the axes.

Secondly. In the hyperbola, let the parallelogram be drawn through the vertices as in the figure.

Put $CG = x'$, $GD = y'$; $CH = x''$, $HE = y''$.

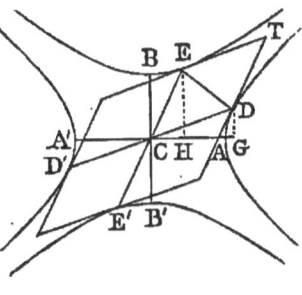

The area of the triangle CDE is equal to the area of the trapezoid HGDE *plus* the area of the triangle CHE *minus* the area of the triangle CGD.

Now the area of the trapezoid HGDE
$\qquad = \tfrac{1}{2}(x' - x'')(y' + y'')$,
the area of the triangle $CHE = \tfrac{1}{2}x''y''$,
the area of the triangle $CDG = \tfrac{1}{2}x'y'$.

But the area of the triangle CDE is equal to one half

of the parallelogram CDTE; hence the area of CDTE
$$= (x'-x'')(y'+y'') + x''y''' - x'y',$$
$$= x'y'' - x''y'.$$
Now $y'' = \dfrac{Bx'}{A}$, and $x'' = \dfrac{Ay'}{B}$;

$$\therefore \text{CDTE} = x'\dfrac{Bx'}{A} - y'\dfrac{Ay'}{B},$$
$$= \dfrac{B^2x'^2 - A^2y'^2}{AB},$$
$$= \dfrac{A^2B^2}{AB} = AB.$$

Hence the parallelogram CDTE is equal to the rectangle described on the semi-axes; and the parallelogram formed by drawing tangents through the vertices of the two conjugate diameters contains four equal parallelograms; therefore it is equal to 4AB, or equal to $2A \times 2B$, as before.

PROPOSITION XII.

(103.) *To find the polar equation of the ellipse or hyperbola.*

The equation of the ellipse or hyperbola, referred to the centre and axes, is
$$A^2y^2 \pm B^2x^2 = \pm A^2B^2, \qquad (1)$$
where the upper signs belong to the ellipse, the lower signs to the hyperbola.

The formulas for passing from rectangular to polar co-ordinates are
$$x = a + r \cos.\theta,$$
$$y = b + r \sin.\theta.$$

Squaring these values of x and y, substituting them in (1), and arranging according to the powers of r, we have

$$\left. \begin{array}{l} A^2 \sin.^2 \theta \\ \pm B^2 \cos.^2 \theta \end{array} \right| r^2 + \left. \begin{array}{l} 2A^2b \sin.\theta \\ \pm 2B^2a \cos.\theta \end{array} \right| r + A^2b^2 \pm B^2a^2 \mp A^2B^2 = 0, \qquad (2)$$

as the general polar equation of the ellipse or hyperbola, the upper signs belonging to the ellipse.

If we place the pole at the right hand focus, the co-ordinates of which are $a = \sqrt{A^2 \mp B^2}$ and $b = 0$, we shall have from (2)
$$(A^2 \sin.^2 \theta \pm B^2 \cos.^2 \theta)r^2 \pm 2B^2a \cos.\theta . r = B^4.$$

Taking the upper signs, dividing by the coefficient of r^2, completing the square, and extracting the root, we obtain

$$r = \frac{-B^2 a \cos.\theta \pm B^2(A^2 \sin.^2\theta + B^2 \cos.^2\theta + a^2 \cos.^2\theta)^{\frac{1}{2}}}{A^2 \sin.^2\theta + B^2 \cos.^2\theta}.$$

Putting $A^2 - B^2$ for a^2 in the above, it reduces to

$$r = \frac{-B^2 a \cos.\theta \pm B^2 A}{A^2 \sin.^2\theta + B^2 \cos.^2\theta} = \frac{-B^2(a \cos.\theta \pm A)}{A^2 \sin.^2\theta + B^2 \cos.^2\theta}.$$

But $\sin.^2\theta = 1 - \cos.^2\theta$. Substituting this in the denominators,

$$r = \frac{-B^2(a \cos.\theta \pm A)}{A^2 - (A^2 - B^2 \cos.^2\theta)} = \frac{-B^2(a \cos.\theta \pm A)}{A^2 - a^2 \cos.^2\theta} =$$

$$\frac{-B^2(a \cos.\theta \pm A)}{(A + a \cos.\theta)(A - a \cos.\theta)};$$

$$r = \frac{-B^2}{A - a \cos.\theta}, \text{ or } r = \frac{B^2}{A + a \cos.\theta}.$$

Now $a = Ae$, and $e^2 = \dfrac{A^2 - B^2}{A^2}$; $\therefore B^2 = A^2(1 - e^2)$. Substituting these in the values for r, we obtain

$$r = \frac{A(1 - e^2)}{1 + e \cos.\theta}, \text{ or } r = -\frac{A(1 - e^2)}{1 - e \cos.\theta}.$$

But $A(1 - e^2)$ is equal to half the parameter of the transverse axis; hence, if we put this equal to p, we shall have

$$r = \frac{p}{1 + e \cos.\theta}, \text{ or } r = \frac{-p}{1 - e \cos.\theta}.$$

(104.) Taking the lower signs for the hyperbola, and proceeding in a manner entirely similar to the above, we shall obtain

$$r = \frac{p}{1 + e \cos.\theta}, \text{ and } r = -\frac{p}{1 - e \cos.\theta},$$

the polar equations of the hyperbola.

Proposition XIII.

(105.) *To find the area of an ellipse.*

Let A'BA be a semi-ellipse, and A'B'A a semicircle described on the transverse axis A'A. Take any point, E, in the ellipse, and draw the ordinate EF, producing

EF, to meet the semicircumference A'B'A in D. Also draw GIH parallel and near to FED, and complete the rectangles GE and GD. Then, by the property of the ellipse,   FD : FE :: A : B.

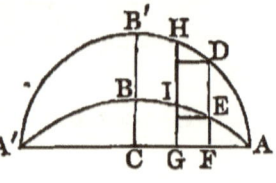

And, Geom., B. IV., Theor. II.,
    the rectangle GD : rectangle GE :: A : B.
Hence, if we suppose a series of rectangles to be inscribed in the semicircle and semiellipse, the same proportion will hold good between each of these, and therefore *the sum of the rectangles in the semicircle will be to the sum of the rectangles in the ellipse as A is to B.* This being true, however great may be the number of these rectangles, is true in the limit.

Now the semicircle is the limit of the rectangles inscribed in it, and the semiellipse is the limit of the rectangles inscribed in it. Therefore
    semicircle : semiellipse :: A : B,
or   area of circle : area of ellipse :: A : B;
    $\therefore$ A × area of ellipse = B × area of circle.
But the area of the circle = πA^2; $\because$ radius = A;
    $\therefore$ area of ellipse = $\dfrac{\pi A^2 B}{A}$ = πAB.

Hence *the area of an ellipse is equal to the product of its semiaxes multiplied by the circumference of a circle whose diameter is unity.*

(106.) THE HYPERBOLA REFERRED TO ITS ASYMPTOTES.

If the diagonals of the rectangle which may be described on the axes of the hyperbola be produced indefinitely, they become the *asymptotes* of the hyperbola.

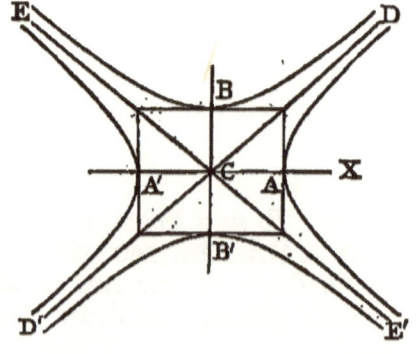

In the accompanying diagram, DD' and EE', diagonals of the rectangle described on the axes AA', BB', indefinitely produced, are the asymptotes.

If we put the angle DCX$=a$, and the angle E'CX$=a'$, we shall then have

$$\frac{B}{A}=\frac{\sin. a}{\cos. a}, \qquad (1)$$

$$\frac{B}{A}=-\frac{\sin. a'}{\cos. a'}, \qquad (2)$$

$$\therefore \frac{\sin.^2 a}{\cos.^2 a}=\frac{B^2}{A^2}, \qquad (3)$$

and
$$\frac{\sin.^2 a}{1-\sin.^2 a}=\frac{B^2}{A^2}; \qquad (4)$$

or $\quad A^2 \sin.^2 a = B^2 - B^2 \sin.^2 a;$

whence $\quad \sin^2 a = \dfrac{B^2}{A^2+B^2};$

$$\therefore \sin. a = \pm \frac{B}{\sqrt{A^2+B^2}}.$$

And from (3),
$$\frac{1-\cos.^2 a}{\cos.^2 a}=\frac{B^2}{A^2},$$

$$A^2 - A^2 \cos.^2 a = B^2 \cos.^2 a;$$

$$\therefore \cos.^2 a = \frac{A^2}{A^2+B^2};$$

$$\therefore \cos. a = \frac{\pm A}{\sqrt{A^2+B^2}}.$$

If $A=B$, the hyperbola is *equilateral*, and we shall then have sin. $a = -$ cos. a, and sin. $a' =$ cos. a'.

(107.) Hence, *in the equilateral hyperbola, the asymptotes are at right angles to each other.*

Proposition XIV.

(108.) *To find the equation of the hyperbola referred to its centre and asymptotes.*

The equation of the hyperbola referred to its centre and axes is

$$A^2 y^2 - B^2 x^2 = -A^2 B^2. \qquad (1)$$

The formulas for passing from rectangular to oblique co-ordinates, the origin remaining the same, are

$$x = x_, \cos. a + y_, \cos. a',$$
$$y = x_, \sin. a + y_, \sin. a'.$$

D

But, since $a = -a'$, these equations become
$$x = (x_{,} + y_{,}) \cos. a,$$
$$y = (x_{,} - y_{,}) \sin. a.$$
Squaring and substituting these in (1), we obtain
$$A^2(x_{,} - y_{,})^2 \sin.^2 a - B^2(x_{,} + y_{,})^2 \cos.^2 a = -A^2 B^2. \quad (2)$$

But $\quad\quad\quad \sin.^2 a = \dfrac{B^2}{A^2 + B^2},$

and $\quad\quad\quad \cos.^2 a = \dfrac{A^2}{A^2 + B^2}.$

Substituting these in (2), we have
$$\frac{A^2 B^2}{A^2 + B^2}(x_{,} - y_{,})^2 - \frac{A^2 B^2}{A^2 + B^2}(x_{,} + y_{,})^2 = -A^2 B^2,$$

or $\quad\quad \dfrac{(x_{,} - y_{,})^2}{A^2 + B^2} - \dfrac{(x_{,} + y_{,})^2}{A^2 + B^2} = -1.$

Clearing of fractions and reducing, we have
$$x_{,} y_{,} = \frac{A^2 + B^2}{4};$$

or, without the *subs.*, $xy = \dfrac{A^2 + B^2}{4},$

which is the equation of the hyperbola referred to its centre and asymptotes.

Since $\dfrac{A^2 + B^2}{4}$ is a constant quantity, we may represent it by the letter m; then
$$xy = m, \quad\quad\quad\quad (3)$$
$$y = \frac{m}{x}.$$

(109.) Now, since m is constant, if x increases continually, y will decrease, and when x becomes infinite, y will become *zero;* hence *the curve continually approaches the asymptote, and the asymptote becomes a tangent at an infinite distance from the origin.*

Proposition XV.

(110.) *To find the equation of a tangent line to a hyperbola referred to its centre and asymptotes.*

The equation of a secant line passing through two points is
$$y - y'' = \frac{y'' - y'}{x'' - x'}(x - x''), \quad\quad (1)$$

THE HYPERBOLA REFERRED TO ITS ASYMPTOTES. 75

And the equation of the hyperbola referred to its centre and asymptotes is

$$xy = m; \quad (2)$$
$$\therefore x'y' = m, \quad (3)$$
$$x''y'' = m; \quad (4)$$
$$\therefore x''y'' - x'y' = 0.$$

Add $x'y''$ to both members; then

$$x''y'' - x'y' + x'y'' = x'y'',$$
or $\quad x'(y'' - y') + y''(x'' - x') = 0;$

$$\frac{y'' - y'}{x'' - x'} = -\frac{y''}{x'}.$$

Substitute this in (1), we shall have

$$y - y'' = -\frac{y''}{x'}(x - x'').$$

If we now suppose the secant line to become a tangent, $x' = x''$, and $y' = y''$, the above will become

$$y - y'' = -\frac{y''}{x''}(x - x''), \quad (5)$$

which is the equation of the tangent line required.

(111.) To find the point at which the tangent line meets the axis of x, make $y = 0$ in (5); we shall then have

$$-y'' = -\frac{y''}{x''}(x - x''),$$

or $\quad x = 2x''$.

That is, the abscissa of the point at which the tangent line cuts the axis is double the abscissa of the point of tangency; $\therefore$ T'M = MC.

Now, since the triangles T'CT and T'MP are similar, we have

$$T'M : MC :: T'P : PT;$$
but $\quad T'M = MC; \therefore T'P = PT.$

Hence, *if a tangent line be drawn through any point of a hyperbola, the part included between the asymptotes is bisected at the point of tangency.*

PROPOSITION XVI.

(112.) *If a tangent line be drawn to the hyperbola, and limited by the asymptotes, it will be equal to the conjugate of the diameter which passes through the point of contact.*

Let the tangent T″T pass through the point P of the hyperbola. Through P draw the semidiameter CP. Put $CP = A'$, and the angle $TCM = \beta$; then, in the triangle CMP, we shall have, by Theor. V., Trig.,

$$-\cos.\ \beta = \frac{x^2 + y^2 - CP^2}{2xy},$$

or $\quad CP^2 = x^2 + y^2 + 2xy \cos.\ \beta;\quad\quad$ (1)

and in the triangle T″MP,

$$\cos.\ \beta = \frac{x^2 + y^2 - PT''^2}{2xy},$$

or $\quad PT''^2 = x^2 + y^2 - 2xy \cos.\ \beta;\quad$ (2)
$\therefore CP^2 - PT''^2 = 4xy \cos.\ \beta.\quad$ (3)

But as the angle $\beta = 2a$, the cos. $\beta = \cos.^2 a - \sin.^2 a$, Trig. (20). And (Art. 106)

$$\cos.\ \beta = \cos.^2 a - \sin.^2 a = \frac{A^2}{A^2 + B^2} - \frac{B^2}{A^2 + B^2} = \frac{A^2 - B^2}{A^2 + B^2}.$$

Also, from the equation of the hyperbola referred to its asymptotes,

$$4xy = A^2 + B^2.$$

Substituting these values of $4xy$ and cos. β in (3), we have

$$CP^2 - PT''^2 = A^2 - B^2 = A'^2 - B'^2 \text{ (Art. 100)}.$$

But $CP = A'$, and hence $PT' = B'$, and $\therefore$ the tangent TT′ is equal to the diameter which is conjugate to P′P.

(113.) Let P′P, E′E be two conjugate diameters, and through their vertices let tangent lines be drawn, forming a parallelogram, $tTT't'$. Then, since the tangents $t't$ and T″T are equal and parallel to the diameter E′E, and the tangents tT, $t'T'$ are equal and parallel to P′P, the vertices of the parallelogram will fall on the asymptotes.

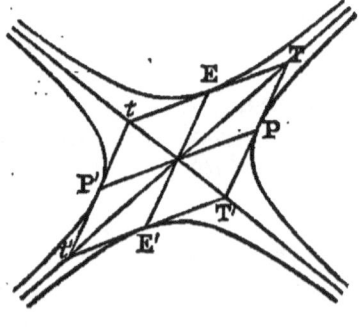

(114.) Hence *the asymptotes are the diagonals of all the parallelograms which can be formed by drawing tangent lines through the vertices of conjugate diameters.*

PROBLEMS.

1. Given the base $=2b$, and the sum of the squares of the other two sides of a triangle $=s^2$, to find the equation of the curve in which the vertex of the triangle is always found.

Let OX and OY be rectangular axes, having their origin at the middle of the base of the triangle; C the vertex of the triangle. Let fall the perpendicular CD.

Put $OD=x$, $DC=y$.
Then $\overline{AC}^2=(b+x)^2+y^2$,
$\overline{BC}^2=(b-x)^2+y^2$;
$\therefore \overline{AC}^2+\overline{BC}^2=2b^2+2x^2+2y^2$,
which, by the problem, is equal to s^2.

Equating and reducing, we have
$$y^2+x^2=\frac{s^2-2b^2}{2},$$
which is the equation of a circle whose centre is O, and radius is $\sqrt{\dfrac{s^2-2b^2}{2}}$.

2. Given the base $=2b$, and the ratio of the other two sides of a triangle as m is to n, to find the equation of the curve in which the vertex is always found.

Using the same figure and co-ordinates as in the preceding problem, we shall have
$\overline{AC}^2=y^2+(b+x)^2$, and $\overline{BC}^2=y^2+(b-x)^2$;
then, Geom., B. III., Theor. IX.,
$$y^2+(b+x)^2 : y^2+(b-x)^2 :: m^2 : n^2;$$
which, by reduction, becomes
$$y^2+x^2-\frac{m^2+n^2}{m^2-n^2}\cdot 2bx+b^2=0,$$
the equation of a circle the co-ordinates of whose centre are $y'=0$, and $x'=\dfrac{m^2+n^2}{m^2-n^2}b$,

and radius $= \sqrt{\left(\dfrac{m^2+n^2}{m^2-n^2} \cdot b\right)^2 - b^2}$.

3. Given the base and vertical angle of a triangle, to find the equation of the curve in which the vertex is always posited.

Using the same figure, let $b=$the base, $v=$the vertical angle, $\phi=$the angle ACD, and $\theta=$the angle BCD. Then
$$v = \phi + \theta.$$
Put $\quad$ OD$=x$, DC$=y$.

Now $\tan. v = \tan.(\phi+\theta) = \dfrac{\tan.\phi + \tan.\theta}{1 - \tan.\phi \tan.\theta}$.

But $\dfrac{DB}{DC} = \tan.\theta$, and $\dfrac{AD}{DC} = \tan.\phi$, and DB$=\tfrac{1}{2}b-x$, and AD$=\tfrac{1}{2}b+x$.

$$\therefore \dfrac{DB}{DC} + \dfrac{AD}{DC} = \dfrac{b}{y} = \tan.\phi + \tan.\theta;$$

$$\therefore \tan. v = \dfrac{\dfrac{b}{y}}{1 - \dfrac{\tfrac{1}{2}b+x}{y} \cdot \dfrac{\tfrac{1}{2}b-x}{y}} = \dfrac{by}{y^2 + x^2 - \tfrac{1}{4}b^2},$$

or $\quad y^2 + x^2 - \tfrac{1}{4}b^2 = \dfrac{by}{\tan. v} = by \cot. v;$

$\therefore y^2 + x^2 - by \cot. v - \tfrac{1}{4}b^2 = 0$,

the equation of a circle the co-ordinates of whose centre are $\quad x'=0$, and $y'=\tfrac{1}{2}b \cot. v$, and the radius $\quad = \tfrac{1}{2}b \operatorname{cosec}. v$.

GENERAL EQUATION OF THE SECOND DEGREE, CONTAINING TWO VARIABLES.

(A.) We shall show in this article that every equation of the second degree containing two variables may be represented geometrically by a point, a straight line, two straight lines, a circle, a parabola, or hyperbola.

The general equation of the second degree may be expressed by
$$Ay^2 + Bxy + Cx^2 + Dy + Ex + F = 0. \qquad (1)$$

And we may consider the axes of co-ordinates to be at right angles to each other; for if we were to pass from

GENERAL EQUATION OF THE SECOND DEGREE. 79

oblique to rectangular, it would not affect the degree of the equation.

If the curve pass through the origin, then $F=0$; but if it do not pass through the origin, F is not equal to zero.

Let us pass from the system of rectangular axes to another system also rectangular, by using the formulas in Art. 21:

$x = x_{,} \cos. a - y_{,} \sin. a$, and $y = x_{,} \sin. a + y_{,} \cos. a$.

Substituting these in (1), we have

$$y_{,}^2 \left\{ \begin{array}{l} A \cos.^2 a \\ -B \sin. a \cos. a \\ +C \sin.^2 a \end{array} \right\} + x_{,}y_{,} \left\{ \begin{array}{l} 2(A-C) \sin. a \cos. a \\ +B (\cos.^2 a - \sin.^2 a) \end{array} \right\}$$

$$+ x_{,}^2 \left\{ \begin{array}{l} A \sin.^2 a \\ +B \sin. a \cos. a \\ +C \cos.^2 a \end{array} \right\} + y_{,} \left\{ \begin{array}{l} D \cos. a \\ -E \sin. a \end{array} \right\}$$

$$+ x_{,} \left\{ \begin{array}{l} D \sin. a \\ +E \cos. a \end{array} \right\} + F = 0. \qquad (2)$$

We can give to a such a value as will render the co-efficient of $x_{,}y_{,}$ equal to zero. Thus,

$2(A-C) \sin. a \cos. a + B (\cos.^2 a - \sin.^2 a) = 0$;

Trig. (18) and (20),

$(A-C) \sin. 2a + B \cos. 2a = 0$;

$\therefore \tan. 2a = -\dfrac{B}{A-C}$.

Hence, if we make a of such a value that

$$\tan. 2a = -\dfrac{B}{A-C},$$

the term containing $x_{,}y_{,}$ will vanish, and we shall have

$$A'y^2 + C'x^2 + D'y + E'x + F = 0. \qquad (3)$$

We may simplify this farther by transferring the origin to a point whose co-ordinates are a and b, by putting

$x = a + x_{,}$, and $y = b + y_{,}$.

Then, substituting these in (3), we shall have

$$A'y_{,}^2 + C'x_{,}^2 + (2A'b + D')y_{,} + (2C'a + E')x_{,} + \left. \begin{array}{l} A'b^2 \\ +C'a^2 \\ +D'b \\ +E'a \\ +F \end{array} \right\} = 0. \qquad (4)$$

Now, if we make $\quad 2A'b + D' = 0$,
and $\qquad\qquad\qquad 2C'a + E' = 0$,

the terms containing x_i and y_i will vanish, and we shall have
$$a = -\frac{E'}{2C'},$$
$$b = -\frac{D'}{2A'}.$$

Using these values of a and b, and putting
$$R = A'b^2 + C'a^2 + D'b + E'a + F,$$
equation (4) will become of the form
$$Py^2 + Qx^2 = -R. \qquad (5)$$

1°. If P, Q, and R have the same sign, the locus is impossible.

2°. If P and Q have the same sign, and R an opposite sign, the curve is an *ellipse*, and the semiaxes are
$$\sqrt{\frac{-R}{Q}} \text{ and } \sqrt{\frac{-R}{P}}.$$

3°. If $P = Q$, the curve is a *circle*.

★ 4°. If P and Q have different signs, the curve is an *hyperbola*.

All these remarks are made under the supposition that R is not equal to zero.

5°. If $R = 0$, and P and Q have the same sign, each of the variables must be equal to zero, and the equation characterizes a point, and that point is the origin of coordinates.

6°. If P and Q have different signs, and $R = 0$, it characterizes two straight lines, represented by the equation
$$y = \pm \sqrt{\frac{-Q}{P}} \cdot x.$$

(B.) If in (3) C' was zero, or, which is the same thing, if the term containing C' were wanting, we should have
$$R = A'b^2 + D'b + E'a + F;$$
and, making $R = 0$, we would find
$$a = -\frac{A'b^2 + D'b + F}{E'};$$
and we also have $b = -\dfrac{D'}{2A'}.$

Substituting these in equation (3), it would become of the form

$$A'y^2 + E'x = 0,$$
or $$y^2 = -\frac{E'}{A'}x,$$
the equation of a *parabola*.

(C.) Because the equation of the circle when referred to the extremity of a diameter is
$$y^2 = 2Rx - x^2,$$
and that of the parabola referred to the vertex is
$$y^2 = 2p,$$
and the ellipse $$y^2 = \frac{B^2}{A^2}(2Ax - x^2),$$

and the hyperbola $$y^2 = \frac{B^2}{A^2}(2Ax + x^2),$$

we may put one general equation to characterize them all. Thus $$y^2 = mx + nx^2,$$
in which m is the parameter of the curve, and n the square of the ratio of the semiaxes.

We might, at first, have transferred the origin to a point whose co-ordinates are a and b, and then assigned such values to a and b as would have caused the coefficients of x, and y, to vanish. In that case we would have found that
$$a = \frac{2AE - BD}{B^2 - 4AC},$$
$$b = \frac{2CD - BE}{B^2 - 4AC}.$$

We shall see that the curves represented by (1) may be divided into two classes,

1°. Those which have a centre, and
2°. Those which generally do not have a centre.

In the first case, the value of $B^2 - 4AC$ does not become equal to zero, but in the second case it does. For if $B^2 - 4AC = 0$, then $a = \infty$ and $b = \infty$.

Removing the terms which have vanished, (1) reduces to $$Ay^2 + Bxy + Cx^2 + R = 0, \qquad (2)$$
where $R = Ab^2 + Bab + Ca^2 + Db + Ea + F$.

Now it is evident that if (2) is satisfied by any *positive* values of x and y, it is also satisfied by the same values taken *negatively*.

Hence *the new origin of co-ordinates must be at the centre of the curve*.

D 2

CHAPTER III.

ANALYTICAL GEOMETRY OF THREE DIMENSIONS.

(115.) We have seen that the position of a point on a plane is determined when its distances from two straight lines drawn in that plane are known; in a manner very similar, we shall now proceed to show that the position of a point *in space* will be determined when its distances from three planes are known.

(116.) Take the three planes YAX, XAZ, YAZ, which are supposed to be at right angles with each other, and whose intersections are the three straight lines AX, AY, and AZ. These three straight lines are each perpendicular to the other two. 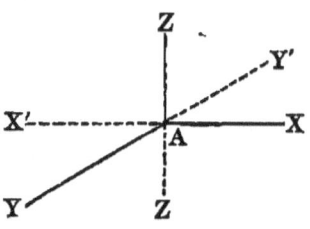 The plane YAX, called the plane *yx*, is supposed to be horizontal, or the plane of the floor on which the student is standing; the planes XAZ and YAZ, called the plane *xz* and *yz* respectively, are both vertical, and may be represented by any two adjacent walls of the room. A is the origin, and the three lines AX, AY, and AZ are the *co-ordinate axes,* called *the axis of x, the axis of y,* and *the axis of z* respectively. The *co-ordinate planes* are indefinite, and hence they will divide all space into eight equal parts or triedral angles, each having its vertex at A. Four of these angles are above the horizontal plane YAX, and four below it. Hence it is necessary, when we obtain the equation of a point in space, to express analytically in which of these eight angles the point is situated. In order to do this, we must apply the principles laid down in the Algebra, Art. 3, which we also applied to distances from points and straight lines in Analytical Geometry of two dimensions: that is, *if distances reckoned along AX to the right of A be considered positive, distances along the same line to the left of*

A *must be regarded as negative.* The same principle must govern in the two other co-ordinate axes.

(117.) The eight triedral angles are numbered as follows:

YAX is called the 1st angle.
YAX'   "   2d   "
X'AY'   "   3d   "
Y'AX   "   4th "

The *fifth* angle is directly under the first, the *sixth* under the second, the *seventh* under the third, and the *eighth* under the fourth.

(118.) If we call the distances of a point in space from the three co-ordinate planes x', y', z', and suppose these distances known, then the position of the point will be completely determined, provided that we have previously ascertained in what triedral angle the point is situated. Let us suppose that the point is in the first angle.

On the three co-ordinate axes take the distances $AD = x'$, $AF = y'$, $AB = z'$, and through the points D, F, B pass planes parallel to the co-ordinate planes. Then, since the plane DG has all its points situated at the distance x' 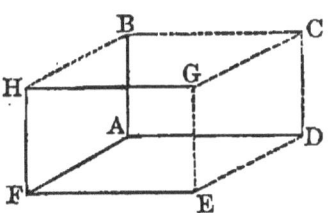 from the plane YAZ, and the plane FG has all its points at a distance y' from the plane XAZ, it follows that all the points of the straight line GE, which is the common intersection of these two planes, have exclusively the property of being at the same distances from the planes YAZ, XAZ. Therefore the point to be determined must be situated in the straight line GE.

Again, the point sought must be situated somewhere in the third plane GHBC, which is parallel to XAY, since all the points in this plane have exclusively the property of being at the distance z' from the plane XAY. Hence the point to be determined must be the point G, in which the third plane cuts the common intersection of the first two, and thus its position is entirely determined. The co-ordinates of the point G are AD, DE, EG.

(119.) If the point be in the first angle, we shall then have for its equations

in the 2d angle, $x=x'$, $y=y'$, $z=z'$;
" 3d " $x=-x'$, $y=y'$, $z=z'$;
" 3d " $x=-x'$, $y=-y'$, $z=z'$;
" 4th " $x=x'$, $y=-y'$, $z=z'$;
" 5th " $x=x'$, $y=y'$, $z=-z'$;
" 6th " $x=-x'$, $y=y'$, $z=-z'$;
" 7th " $x=-x'$, $y=-y'$, $z=-z'$;
" 8th " $x=x'$, $y=-y'$, $z=-z'$.

(120.) There are also some particular positions of the point which it is proper to notice. For example, in order to express that a point is situated in the plane xy, we must make its distance from that plane *zero*; then we shall have for the equations of a point in the plane xy,
$$x=x', y=y', z=0.$$
Similarly, a point on the axis of x,
$$x=x', y=0, z=0.$$

(121.) *If from any point in space a straight line be drawn perpendicular to a given plane, the foot of the perpendicular is called the projection of the given point upon the given plane.*

(122.) *In like manner, if from every point of any line in space, whether straight or curved, perpendiculars be drawn to any given plane, the line traced out by the feet of the perpendiculars upon the given plane is called the projection of the given line upon the given plane.*

(123.) If we say that $x=x'$, $y=y'$, $z=z'$ are the equations of the point G, then the co-ordinates of the point E are
$$x=x', y=y', z=0. \qquad (1)$$
The co-ordinates of the point C are
$$x=x', y=0, z=z', \qquad (2).$$
which give for the co-ordinates of the point H
$$x=0, y=y', z=z'. \qquad (3)$$
Hence, if the projection of a point G upon two of the co-ordinate planes be given, the third projection will also necessarily be known.

(124.) When the co-ordinate planes are not at right angles to each other, the axes AX, AY, AZ make oblique angles with each other, and are called *oblique axes;* the equations of a point, however, are $x=x'$, $y=y'$, $z=z'$; but x', y', and z' express distances reckoned along lines respectively parallel to these oblique axes.

ANALYTICAL GEOMETRY OF THREE DIMENSIONS.   85

In other respects, the remarks made with regard to rectangular axes are applicable to oblique axes also.

PROPOSITION I. PROBLEM.

(125.) *To find the distance between two points in space when the co-ordinates of these points are known.*

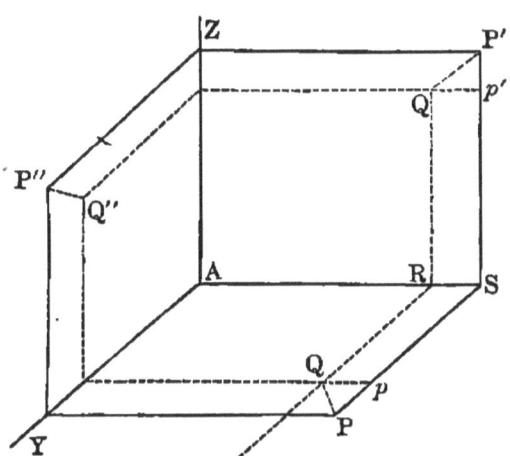

Let Q, Q', Q'' be the projections of one of the points on the three co-ordinate planes, and P, P', P'' be the projections of the other point.

Then $AR=x'$, $RQ=y'$, $RQ'=z'$ are the co-ordinates of the point Q, and $AS=x''$, $SP=y''$, and $SP'=z''$ the co-ordinates of the point P.

Now, in the triangle QpP, right-angled at p, we have
$$QP^2 = Qp^2 + pP^2 = (x''-x')^2 + (y''-y')^2.$$
But QP is the base of a right-angled triangle whose perpendicular is $p'P'$, and hypothenuse the line whose length is required. Putting D equal to the length of this line, we shall have
$$D^2 = (x''-x')^2 + (y''-y')^2 + (z''-z')^2;$$
$$\therefore D = \sqrt{(x''-x')^2 + (y''-y')^2 + (z''-z')^2}.$$

(126.) If the point Q be at the origin, its co-ordinates become each equal to zero, and the above expression becomes
$$D = \sqrt{x''^2 + y''^2 + z''^2}. \tag{A}$$

(127.) From which it appears that *the square of the*

diagonal of a rectangular parallelopipedon is equal to the sum of the squares of the three edges.

(128.) From this we may easily determine a relation which exists between the cosines of the angles which a straight line makes with the co-ordinate axes.

Thus, let $r=$ the line passing through the origin; X, Y, Z the angles which this line makes with the axes of x, y, and z respectively. Then the co-ordinates of the other extremity of this line being x'', y'', z'', we shall have

$$x''=r \cos. X, \quad y''=r. \cos. Y, \quad z''=r. \cos. Z.$$

Squaring these equations and adding them together, we have

$$x''^2+y''^2+z''^2=r^2(\cos.^2 X+\cos.^2 Y+\cos.^2 Z).$$

But from (A) we have the square of the length of a line equal to $x''^2+y''^2+z''^2$, and, as we represented this length by r, therefore

$$r^2=r^2(\cos.^2 X+\cos.^2 Y+\cos.^2 Z),$$
or $\quad\cos.^2 X+\cos.^2 Y+\cos.^2 Z=1.$

PROPOSITION II.

To find the equations of a straight line in space.

(129.) The projections of a straight line on two planes are sufficient to determine its position, and hence it follows that a straight line in space will be determined analytically if we know the equations of its projections upon two of the three co-ordinate planes.

We generally consider the projections of the straight line on the planes XZ and YZ; and since these two planes have AZ for their common axis, this line is regarded in each of the planes as the axis of abscissas. Therefore AX is the axis of ordinates in the plane XZ, and AY is the axis of ordinates in the plane YZ.

Let DE be the projection of a straight line on the co-ordinate plane XZ, and KG its projection on the plane YZ. Through the origin A draw AP and AL respectively parallel to DE and KG.

Put $BP=x$, $AB=z$, and $a=$ the tangent of the angle EHZ, or PAB.

Then, in the triangle BAP, we have

$$\frac{BP}{AB}=\tan. BAP,$$

ANALYTICAL GEOMETRY OF THREE DIMENSIONS. 87

or $\quad \dfrac{x}{z}=a;$

$\therefore x=az.$

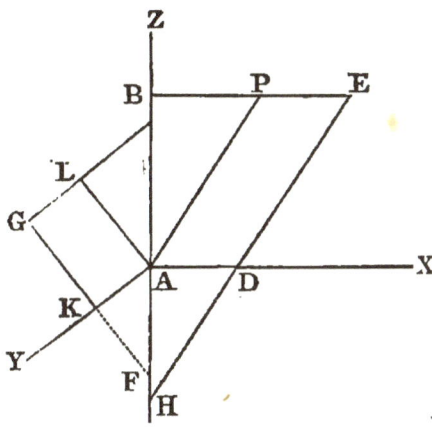

But the point E in the line ED is farther from the axis of Z than P is by the distance PE or AD, which may be called a; hence the equation of the line ED becomes

$$x=az+a. \quad (1)$$

In a manner entirely similar, we find

$$y=bz+\beta \quad (2)$$

for the equation of the line GK.

These, being the equations of the projections of the line on the co-ordinate planes XZ and YZ, are called *the equations of the line in space*.

Now the equation $x=az+a$ expresses not only the relation between the co-ordinates of any point in the line DE, but it also expresses the relation between the co-ordinates of any point in the plane drawn through the line in space perpendicular to the plane XZ.

In like manner, the equation

$$y=bz+\beta$$

belongs not only to the straight line GK, but also to all the points of the plane drawn through the line in space perpendicular to the plane YZ. This system of equations, therefore, holds good for all the points of the straight line in space, which is the intersection of the two planes perpendicular to the planes XZ and YZ, and holds good for the points of this straight line alone. Therefore these equations are the equations of the straight line itself.

If we eliminate z from these equations, we shall have

$$y-\beta=\dfrac{b}{a}(x-a) \quad (3)$$

for the equation of the projection on the plane XY.

(130.) When the straight line passes through the origin, the equations which we obtained first, viz.,

$$x = az,$$
$$y = bz,$$
are the equations of that line.

(131.) If the straight line be situated in one of the co-ordinate planes, as, for instance, in the plane xz, we shall have $y=0$, and $x=az+\alpha$.
That is, in this case, $b=0$, and $\beta=0$, which is evident, for the projection of the straight line on the plane yz will coincide with AZ.

(132.) When the constants a, b, α, β are given, the position of the straight line is completely determined. In order to obtain its different points, we must give a succession of particular values to one of the variables in each of the equations
$$x = az + \alpha,$$
$$y = bz + \beta,$$
by means of which we shall obtain corresponding values for the two other variables.

PROPOSITION III.

(133.) *To find the equations of a straight line in space which passes through one given point, and then for a straight line through two given points.*

Let the co-ordinates of the point be x', y', z'. The equations to the straight line will be of the form
$$x = az + \alpha, \qquad (1)$$
$$y = bz + \beta. \qquad (2)$$
Now, since the line passes through this point, we shall have
$$x' = az' + \alpha, \qquad (3)$$
$$y' = bz' + \beta. \qquad (4)$$
And, since these equations hold good together for the straight line required, we shall have
$$x - x' = a(z - z'), \qquad (5)$$
$$y - y' = b(z - z'), \qquad (6)$$
the equations of a line in space which passes through a given point.

Again, let the co-ordinates of one given point be x', y', z', and the co-ordinates of the other given point be x'', y'', z''. Now, since it passes through this second point, we must have

$$x'' = az'' + a, \qquad (7)$$
$$y'' = bz'' + \beta. \qquad (8)$$
Subtracting (7) from (3) and (8) from (4), we obtain
$$x' - x'' = a(z' - z'');$$
$$\therefore a = \frac{x' - x''}{z' - z''}. \qquad (9)$$
$$y' - y'' = b(z' - z'');$$
$$\therefore b = \frac{y' - y''}{z' - z''}. \qquad (10)$$
Substituting these values of a and b in (5) and (6), we have
$$x - x' = \frac{x' - x''}{z' - z''}(z - z'), \qquad (11)$$
and
$$y - y' = \frac{y' - y''}{z' - z''}(z - z'), \qquad (12)$$
which are the equations of a straight line in space passing through two given points.

PROPOSITION IV.

(134.) *Through a given point without a straight line in space, to draw a straight line parallel to the given line.*
Let the equations of the given straight line be
$$x = az + a,$$
$$y = bz + \beta,$$
and x', y', z' the co-ordinates of the given point. The equations of the required straight line will be of the form
$$x - x' = A(z - z'),$$
$$y - y' = B(z - z'),$$
where A and B are quantities to be determined. Since the straight lines are to be parallel, the planes which project them respectively upon the planes of xz and yz must also be parallel; hence the intersections of these parallel planes with the co-ordinate planes must be parallel, or, in other words, the projections of the straight lines must be parallel.
$$\therefore A = a, \text{ and } B = b,$$
which give for the equations of the required line
$$x - x' = a(z - z'),$$
$$y - y' = b(z - z').$$

Proposition V.

(135.) *Given the equations of two straight lines in space, to determine the relation which must exist between the constants, in order that the two lines may intersect, and to find the co-ordinates of the point of intersection.*

Let
$$x = az + \alpha, \quad (1)$$
$$y = bz + \beta, \quad (2)$$
$$x = a'z + \alpha', \quad (3)$$
$$y = b'z + \beta', \quad (4)$$

be the equations of the two given straight lines.

If these straight lines intersect each other in space, their equations must hold good together at the point of intersection. Therefore, eliminating x, y, z, we find the relation

$$(\alpha - \alpha')(b - b') = (\beta - \beta')(a - a');$$

and, unless this equation of condition be satisfied, the lines do not intersect; if it holds good, then the point of intersection has for its co-ordinates

$$z = \frac{\alpha' - \alpha}{a - a'} = \frac{\beta' - \beta}{b - b'}, \text{ and } y = \frac{b\beta' - b'\beta}{b - b'},$$

and
$$x = \frac{a\alpha' - a'\alpha}{a - a'}.$$

Proposition VI.

(136.) *Given the equations of two straight lines in space which intersect, to find the angle contained between them.*

Let the equations of the given lines be

$$x = az + \alpha, \quad y = bz + \beta, \quad (1)$$
$$x = a_,z + \alpha_,, \quad y = b_,z + \beta_,, \quad (2)$$

A being the origin of co-ordinates. Through A draw two straight lines, AP′, AP″ respectively, parallel to the given straight lines.

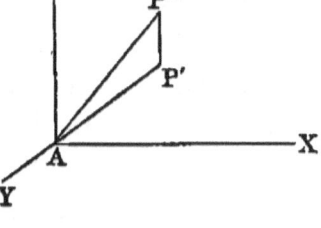

Then the angle P′AP″ will be equal to the required angle.

Put $\angle \text{P}'\text{AP}'' = V.$

The equations of AP′ are $x = az$, $y = bz$, (3)

The equations of AP'' are $x = a_{,}z$, $y = b_{,}z$. (4)

In AP' take *any point*, P', and let its co-ordinates be represented by $x_{,}, y_{,}, z_{,}$; and in AP'' take any point, P'', and represent its co-ordinates by x_2, y_2, z_2.

These co-ordinates must satisfy (3) and (4).

$$\therefore x_{,} = az_{,}, \quad y_{,} = bz_{,}, \quad (5)$$
$$x_2 = a_{,}z_2, \quad y_2 = b_{,}z_2. \quad (6)$$

Join $P'P''$, and let $AP' = r$, $AP'' = r_{,}$.

Then, Trig., Theor. V., $\cos. V = \dfrac{r^2 + r_{,}^2 - \overline{P'P''}^2}{2rr_{,}}$;

$$\therefore \overline{P'P''}^2 = r^2 + r_{,}^2 - 2rr_{,} \cos. V.$$

But $\overline{P'P''}^2 = (x_{,} - x_2)^2 + (y_{,} - y_2)^2 + (z_{,} - z_2)^2$.

Equating these two values of $\overline{P'P''}^2$, and remembering that $x_{,}^2 + y_{,}^2 + z_{,}^2 = r^2$, and $x_2^2 + y_2^2 + z_2^2 = r_{,}^2$, we shall have

$$\cos. V = \frac{x_{,}x_2 + y_{,}y_2 + z_{,}z_2}{rr_{,}}. \quad (7)$$

From (5) and (6) we have

$$x_{,}^2 = a^2 z_{,}^2, \quad y_{,}^2 = b^2 z_{,}^2; \quad x_2^2 = a_{,}^2 z_2^2, \quad y_2^2 = b_{,}^2 z_2^2;$$
and $\quad x_{,}x_2 = aa_{,}z_{,}z_2, \quad y_{,}y_2 = bb_{,}z_{,}z_2$.

Adding, we have

$$x_{,}^2 + y_{,}^2 + z_{,}^2 = z_{,}^2(a^2 + b^2 + 1) = r^2,$$
$$x_2^2 + y_2^2 + z_2^2 = z_2^2(a_{,}^2 + b_{,}^2 + 1) = r_{,}^2;$$

$\therefore r = z_{,}\sqrt{1 + a^2 + b^2}$, and $r_{,} = z_2\sqrt{1 + a_{,}^2 + b_{,}^2}$.

Substituting these in (7), we obtain

$$\cos. V = \frac{1 + aa_{,} + bb_{,}}{\sqrt{1 + a^2 + b^2}\sqrt{1 + a_{,}^2 + b_{,}^2}}. \quad (A)$$

(137.) In Eq. (A), let $V = 90°$; then $\cos. V = 0$, and

$$\therefore 1 + aa_{,} + bb_{,} = 0,$$

which is the equation of condition for two straight lines in space to be perpendicular to each other.

(138.) Again, if we make $V = 0$, $\cos. V = 1$, and the two straight lines become parallel to each other, and equation (A) becomes

$$\frac{1 + aa_{,} + bb_{,}}{\sqrt{1 + a^2 + b^2}\sqrt{1 + a_{,}^2 + b_{,}^2}} = 1,$$

which, being cleared of fractions, squared, and reduced, becomes

$$(a_{,} - a)^2 + (b_{,} - b)^2 + (ab_{,} - a_{,}b)^2 = 0.$$

Since each of the terms is a square, and on that account essentially positive, the equation can only be satisfied when the terms are separately equal to *zero*. These conditions give
$$a_{,}=a, \ b_{,}=b.$$
The projections of the lines on each of the co-ordinate planes are therefore parallel to each other.

Proposition VII.

(139.) *To find the angle which either of these straight lines in space makes with each of the axes of co-ordinates.*

Let the angle which the line AP' makes with the axis of x be denoted by X, the angle which it makes with the axis of y be denoted by Y, and the angle which it makes with the axis of z be denoted by Z.

Then $x = r \cdot \cos X$; $\therefore \cos X = \dfrac{x}{r}$.

$y = r \cdot \cos Y$; $\therefore \cos Y = \dfrac{y}{r}$.

$z = r \cdot \cos Z$; $\therefore \cos Z = \dfrac{z}{r}$.

But $r^2 = x^2 + y^2 + z^2 = z^2(1 + a^2 + b^2)$,

or $r = z\sqrt{1 + a^2 + b^2}$;

$\therefore \cos X = \dfrac{x}{z\sqrt{1 + a^2 + b^2}}$.

But $x = az$;

$\therefore \cos X = \dfrac{a}{\sqrt{1 + a^2 + b^2}}$.

Similarly, $\cos Y = \dfrac{b}{\sqrt{1 + a^2 + b^2}}$,

$\cos Z = \dfrac{1}{\sqrt{1 + a^2 + b^2}}$.

(140.) Since $x^2 + y^2 + z^2 = r^2(\cos^2 X + \cos^2 Y + \cos^2 Z)$, and $x^2 + y^2 + z^2 = r^2$,

$\therefore \cos^2 X + \cos^2 Y + \cos^2 Z = 1$;

which shows that *the sum of the squares of the cosines of the angles which a straight line in space forms with the three axes is equal to unity.*

PROPOSITION VIII.

(141.) *To find the angle included between two lines in space given by their equations in terms of the angles which each of the straight lines makes with the axes of co-ordinates.*

Let the angles which r makes with the axes AX, AY, AZ, be denoted by X, Y, Z, and the angles which $r_{\prime}$ makes with the same axes be denoted by $X_{\prime}, Y_{\prime}, Z_{\prime}$ respectively.

Then, as before,

$$\cos. V = \frac{x_{\prime}x_2 + y_{\prime}y_2 + z_{\prime}z_2}{rr_{\prime}}.$$

But $x_{\prime} = r.\cos. X$, and $x_2 = r_{\prime}\cos. X_{\prime}$,
$y_{\prime} = r.\cos. Y$, $\quad y_2 = r_{\prime}\cos. Y_{\prime}$,
$z_{\prime} = r.\cos. Z$, $\quad z_2 = r_{\prime}\cos. Z_{\prime}$.

∴ $\cos. V = \cos. X \cos. X_{\prime} + \cos. Y \cos. Y_{\prime} + \cos. Z \cos. Z_{\prime}$.

That is

(142.) *The cosine of the angle included between two lines in space is equal to the sum of the rectangles of the cosines of the angles which the lines in space form with the co-ordinate axes.*

OF THE PLANE.

(143.) *The equation of a plane is an equation which expresses the relations between the co-ordinates of every point of the plane.*

PROPOSITION IX.

(144.) *To find the equation of a plane.*

A straight line is said to be perpendicular to a plane when it is perpendicular to all the straight lines in the plane which pass through the point in which it meets the plane. Hence a plane may be generated by the motion of a straight line round a point in another straight line to which it is perpendicular.

Let $\quad x = az + \alpha, \; y = bz + \beta,\quad$ (1)

be the equations of a given line, and
$$x', y', z'$$

the co-ordinates of the point in this line through which the generating line passes. Then its equations will be
$$x-x'=a(z-z'), \quad (2)$$
$$y-y'=b(z-z'). \quad (3)$$
And the equations of the perpendicular passing through the given point will be
$$x-x'=a_{,}(z-z_{,}), \quad (4)$$
$$y-y'=b_{,}(z-z_{,}). \quad (5)$$

From (4) we have $\quad a_{,}=\dfrac{x-x'}{z-z'}.\quad (6)$

From (5) we have $\quad b_{,}=\dfrac{y-y'}{z-z'}.\quad (7)$

But the equation of condition for two lines to be at right angles to each other is
$$1+aa_{,}+bb_{,}=0. \quad (8)$$
Substituting the values of $a_{,}$, $b_{,}$, found in (6) and (7), we obtain
$$1+a\cdot\dfrac{x-x'}{z-z'}+b\cdot\dfrac{y-y'}{z-z'}=0.$$
$$\therefore z+ax+by-(z'+ax'+by')=0.$$
But a, b, x', y', z' are known quantities, and we may therefore represent $z'+ax'+by'$ by a single letter, c. Doing this, we have
$$z+ax+by-c=0.$$

(145.) *The equation of the plane*, which is an equation of *the first degree containing three variables.*

(146.) The lines in which a plane intersects the co-ordinate planes are called *the traces of the plane.*

We can find these traces by combining the equation of the plane with each of the equations of the co-ordinate planes.

At every point in the plane xz, y is equal to zero; if, then, we make $y=0$ in the equation of the plane, we shall have
$$x=-\dfrac{1}{a}z+\dfrac{c}{a}$$
for *the equation of the trace* BC on the plane xz.

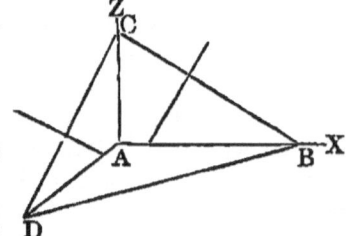

In this equation $-\dfrac{1}{a}$ is the

tangent of the angle which the trace BC makes with the axis of Z, and $\frac{c}{a}$ is the distance AB.

Similarly, $$y = -\frac{1}{b}z + \frac{c}{b},$$

is *the equation of the trace* DC *on the plane yz.*

(147.) Now, when the line which was perpendicular to the generating line was projected on the plane xz, that projection made with the axis of z an angle whose tangent was a; and we find that the trace BC of the plane on the same co-ordinate plane makes with the same axis, z, an angle whose tangent is $-\frac{1}{a}$. Hence the product of these two tangents is equal to *minus unity*, and therefore the projection of the line is perpendicular to the trace. The same may be shown for the other traces.

(148.) Therefore, *if a line in space be perpendicular to a plane, the projections of the line will be respectively perpendicular to the traces of the plane.*

(149.) The most general form of the equation of a plane is
$$Ax + By + Cz + D = 0.$$

PROPOSITION X.

(150.) *To find the equation of a plane which shall pass through three given points.*

Let x', y', z'; x'', y'', z''; and x''', y''', z''', be the co-ordinates of the three given points.

The equation will be of the form
$$z = Ax + By + C. \qquad (1)$$
And, since it passes through a point whose co-ordinates are x', y', z', these co-ordinates must satisfy (Eq. 1).
$$\therefore z' = Ax' + By' + C. \qquad (2)$$
Subtracting (1) from (2),
$$z' - z = A(x - x') + B(y - y'), \qquad (3)$$
which is the equation of a plane passing through *one* given point.

The constants A and B being arbitrary, the problem will be indeterminate, as it should be, for any number of planes may be passed through one given point.

Let the plane pass through the second point whose

co-ordinates are x'', y'', z'', then these must satisfy the equation, and we shall have
$$z''=Ax''+By''+C. \qquad (4)$$
We may eliminate two of the constants by (1), (2), (4). One, however, will still remain, and the problem will be indeterminate, for any number of planes may be passed through *two* given points.

Finally, let the plane pass through the third point also, the co-ordinates being x''', y''', z''', and we shall have
$$z'''=Ax'''+By'''+C. \qquad (5)$$
The three constants may then be eliminated between equations (1), (2), (4), (5), and the result will be the equation of the only plane that can pass through the three given points.

Problems.

1. *Find the equation of a plane which shall pass through the three points* whose co-ordinates are

$x' = 1,\quad y' = -2,\quad z' = 2;$
$x'' = 0,\quad y'' = 4,\quad z'' = -5;$
$x''' = -2,\quad y''' = 1,\quad z''' = 0.$

If we substitute the given co-ordinates of the points in the general equation $z=Ax+By+C$, we shall have the three equations,

$$2 = A-2B+C, \qquad (1)$$
$$-5 = 4B+C, \qquad (2)$$
$$0 = -2A+B+C, \qquad (3)$$

from which we have $A=-\dfrac{3}{5}$, $B=-\dfrac{19}{15}$, $C=\dfrac{1}{15}$.

Substituting these in the general equation, we shall have
$$15z=-9x-19y+1,$$
the equation of the required plane.

2. Find the equation of a plane which shall pass through the three points whose co-ordinates are

$x' = 2,\quad y' = -1,\quad z' = 3;$
$x'' = 3,\quad y'' = -2,\quad z'' = 5;$
$x''' = -3,\quad y''' = 0,\quad z''' = -2.$

3. Find the equation of a plane which shall pass through the three points whose co-ordinates are

$$x' = 2, \quad y' = -1, \quad z' = 3;$$
$$x'' = 3, \quad y'' = -2, \quad z'' = 5;$$
$$x''' = -7, \quad y''' = 4, \quad z''' = -1.$$

PROPOSITION XI.

(151.) *To find the equations of the intersection of two planes.*

Let the equations of the two planes be
$$z = Ax + By + C,$$
$$z = A'x + B'y + C'.$$

Then, since these equations hold good together for the straight line which is their common intersection, if we eliminate z, we shall have the equation of the projection of that intersection on the plane xy; that is,
$$(A - A')x + (B - B')y + C - C' = 0.$$

We may find, in a manner entirely similar, the equation of the projection of the intersection on the co-ordinate plane yz, and also on the co-ordinate plane xz.

PROPOSITION XII.

(152.) *To find the conditions which will cause a straight line and a plane to be parallel or coincide.*

Let the equation of the straight line be
$$x = az + a, \quad y = bz + \beta,$$
and the equation of the plane
$$z = Ax + By + C.$$

In the equation of the plane, substitute for x and y their values $az + a$ and $bz + \beta$, and we shall have
$$(Aa + Bb - 1)z + (Aa + B\beta + C) = 0; \quad (A)$$
$$\therefore z = -\frac{Aa + B\beta + C}{Aa + Bb - 1}.$$

Hence, if the straight line and plane have only one point in common, that is, if the line pierces the plane, the co-ordinates of the common point will satisfy the equations of the line and plane, and we would be able to determine its co-ordinates; but if the straight line be altogether in the given plane, the equation of condition (A) must hold good, *whatever may be the value of z.* Hence the two parts of the equation must be independent of each other, and we shall have

E

$$Aa+Bb-1=0,$$
and $$Aa+B\beta+C=0,$$
the equations of condition required.

If the straight line be merely parallel to the given plane, if we move them in a direction parallel to their original position until they reach the origin, the plane and straight line will coincide; hence the above equations must be satisfied on the supposition that a, and β, and C are each equal to *zero;* therefore
$$Aa+Bb-1=0$$
will be *the equation of condition which must be satisfied in order that a straight line and a plane may be parallel.*

Proposition XIII.

(153.) *To find the conditions which will cause a straight line to be perpendicular to a plane.*

If a straight line be perpendicular to a plane, the projections of the line and the traces of the plane will be respectively perpendicular to each other.

Let the equations of the given plane and straight line be
$$z=Ax+By+C, \quad (1)$$
$$x=az+a, \quad (2)$$
$$y=bz+\beta. \quad (3)$$

Eq. 2 is the equation of the projection of the given straight line on the co-ordinate plane xz, and (3) is the projection on yz.

The trace of the plane on xz is $x=\dfrac{1}{A}z-\dfrac{C}{A}.$ (4)

The trace of the plane on yz is $y=\dfrac{1}{B}z-\dfrac{C}{B}.$ (5)

But, since the projections of the line must be perpendicular to the traces of the plane, we shall have
$$a\times\frac{1}{A}+1=0;$$
$$\therefore A=-a, \text{ and } B=-b,$$
which are the required conditions.

Proposition XIV.

(154.) *To draw from a given point a line perpendicular to a given plane, and to find the length of the perpendicular.*

OF THE PLANE. 99

Let the equation of the plane be
$$z = Ax + By + C, \tag{1}$$
and let x', y', z' be the co-ordinates of the given point. Then the equations of the required line must be of the form
$$x - x' = a(z - z'),$$
$$y - y' = b(z - z').$$
And, since it is perpendicular to the plane,
$$a = -A, \quad b = -B,$$
and therefore the equations of the perpendicular are
$$x - x' + A(z - z') = 0, \tag{2}$$
$$y - y' + B(z - z') = 0. \tag{3}$$

If we denote the co-ordinates of the point in which the line pierces the plane by x'', y'', z'', equations (1), (2), (3) will become
$$z'' = Ax'' + By'' + C, \tag{4}$$
$$x - x'' = -A(z - z''), \tag{5}$$
$$y - y'' = -B(z - z''). \tag{6}$$

Combining (1) and (4), we have
$$z - z'' = A(x - x'') + B(y - y''). \tag{7}$$

If to the second member of (7) we add C and subtract its equal $(z'' - Ax'' - By'')$, it will take the form
$$z - z'' = A(x - x'') + B(y - y'') + C - z'' + Ax'' + By''. \tag{8}$$

Now, equations (5), (6), (8) will hold good together.
$$\therefore z - z'' = -A^2(z - z'') - B^2(z - z'') + C - z'' + Ax'' + By'',$$
or $\quad (z - z'')(1 + A^2 + B^2) = C - z'' + Ax'' + By''.$

$$\therefore z - z'' = \frac{C - z'' + Ax'' + By''}{1 + A^2 + B^2};$$

$$\therefore y - y'' = -\frac{B(C - z'' + Ax'' + By'')}{1 + A^2 + B^2},$$

and $\quad x - x'' = \frac{-A(C - z'' + Ax'' + By'')}{1 + A^2 + B^2}.$

And the distance $D = \sqrt{(x - x'')^2 + (y - y'')^2 + (z - z'')^2}$;
$$\therefore D = \frac{z - Ax'' - By'' - C}{\sqrt{1 + A^2 + B^2}},$$
the distance required.

PROPOSITION XV.

(155.) *To find the angle included between two planes.*
Let the equations of the planes be

$$z = Ax + By + C, \qquad (1)$$
$$z = A_{,}x + B_{,}y + C_{,}. \qquad (2)$$

(156.) If we let fall from the origin two straight lines perpendicular on these planes, the angles contained by the straight lines will be equal to the angle contained by the planes.

Let the equations to these straight lines be
$$\left.\begin{matrix} x = az, \\ y = bz, \end{matrix}\right\} \text{ and } \left\{\begin{matrix} x = a_{,}z, \\ y = b_{,}z. \end{matrix}\right.$$

The angle between them is given by the equation
$$\cos. V = \frac{1 + aa_{,} + bb_{,}}{\sqrt{1 + a^2 + b^2}\sqrt{1 + a_{,}^2 + b_{,}^2}}. \qquad (3)$$

But, in order that the straight lines may be perpendicular to the given planes, we must have
$$A = -a, \quad B = -b, \quad A_{,} = -a_{,}, \quad B_{,} = -b_{,}.$$

Substituting these values, we have the cosine of the angle between the two planes
$$\cos. V = \frac{1 + AA_{,} + BB_{,}}{\sqrt{1 + A^2 + B^2}\sqrt{1 + A_{,}^2 + B_{,}^2}}. \qquad (4)$$

If we wish to find the angle which any plane makes with the co-ordinate planes, we have only to suppose that one of the above planes assumes in succession the position of the different co-ordinate planes: thus, that (Eq. 2) is the plane of xy; then its equation becomes, by making $z = 0$,
$$A_{,}x + B_{,}y = 0.$$

And, since this is true for all values of x and y, we shall have
$$A_{,} = 0, \; B_{,} = 0.$$

Designating the angles formed by the planes by V', we have
$$\cos. V' = \frac{1}{\sqrt{1 + A^2 + B^2}}. \qquad (5)$$

If we designate by V'' and V''' the angles which the given plane makes with the planes xz, yz, we shall obtain
$$\cos. V'' = \frac{B}{\sqrt{1 + A^2 + B^2}}, \qquad (6)$$
$$\cos. V''' = \frac{A}{\sqrt{1 + A^2 + B^2}}. \qquad (7)$$

If we square the three values of cos. V', cos. V", cos. V''', and add them together, we find

$$\cos.^2 V' + \cos.^2 V'' + \cos.^2 V''' = 1.$$

(157.) That is, *the sum of the squares of the cosines of the three angles which a plane forms with the three co-ordinate planes is equal to radius square or unity.*

(158.) If we now suppose the first plane to coincide in succession with each of the co-ordinate planes, and designate by V_1, V_2, V_3 the angles formed by the second plane with the co-ordinate planes, we shall have

$$\cos. V_1 = \frac{1}{\sqrt{1 + A_{\prime}^2 + B_{\prime}^2}}, \qquad (8)$$

$$\cos. V_2 = \frac{B_{\prime}}{\sqrt{1 + A_{\prime}^2 + B_{\prime}^2}}, \qquad (9)$$

$$\cos. V_3 = \frac{A_{\prime}}{\sqrt{1 + A_{\prime}^2 + B_{\prime}^2}}. \qquad (10)$$

If we multiply (5) by (8), (6) by (9), (7) by (10), member by member, add the results, and compare the equation with (4), we shall find that

$$\cos. V = \cos. V' \cos. V_1 + \cos. V'' \cos. V_2 + \cos. V''' \cos. V_3,$$

which expresses the cosine of the angle included between two planes in terms of the angles which the planes form with the co-ordinate axes.

DIFFERENTIAL CALCULUS.

CHAPTER I.
DEFINITIONS AND INTRODUCTORY REMARKS. — DIFFERENTIATION OF ALGEBRAIC FUNCTIONS.—DIFFERENTIAL COEFFICIENT OF A FUNCTION OF A FUNCTION.

1. IN the Differential Calculus we shall meet with quantities which always retain the same values during the same investigation, while others are subject to certain laws of change, by which they may take, in succession, an infinite number of different values without changing the form of the expression into which they enter.

2. The former of these are called *constant quantities*, and are represented by the first letters of the alphabet. The latter are called *variable quantities*, and are represented by the final letters of the alphabet.

3. Some variable quantities are confined *within certain limits*, while others vary *without limit*: thus, the cosine of an arc commencing at *zero* degrees is equal to the radius, but as the arc increases the cosine diminishes; when the arc becomes 90° the cosine is equal to *zero;* and after the arc passes that point the cosine becomes negative, reaching its greatest negative value at 180°, being at that point equal to *minus* the radius. At 270° it again becomes equal to *zero;* and when the arc is more than 270°, the cosine becomes positive, and continues positive till its value becomes *zero* again.

On the other hand, the tangent commencing at *zero* degrees is equal to *zero*, but as the arc increases the tangent increases also. When the arc becomes 90° the tangent becomes *infinite*, and after the arc passes that point the tangent becomes negative, and diminishes in value till the arc becomes 180°: here the tangent is equal to *zero*. Passing this point, the tangent again becomes positive, etc. Hence we see that the cosine varies from *plus radius* to *minus radius*, and the tangent varies from *plus infinity* to *minus infinity*.

4. It is important for the student to understand what is meant by the term *limit*, or *limiting* value, and passing to the *limit*. If a regular polygon be inscribed in a circle, and we inscribe another polygon having double the number of sides, the perimeter of the second polygon will approach more nearly in value to the circumference of the circle than that of the first; and if we continue the process of inscribing polygons, each having double the number of sides of the other, the perimeters of these polygons will approach nearer and nearer to the circumference of the circle, and the last may be made to differ from it by less than any assignable quantity. Hence the circle is said to be the *limit* of the inscribed polygon.

5. If we have the series $\frac{1}{10} + \frac{1}{100} + \frac{1}{1000} +$ etc., to *infinity*, we readily perceive that the sum of the series approaches in value to $\frac{1}{9}$, but can never equal it while the number of terms is *finite*. We therefore say that $\frac{1}{9}$ is the *limit* of that series.

6. If
$$u = ax,$$
the product of a and x becomes less and less as x diminishes in value. Hence
$$u = 0$$
when x becomes equal to zero; or, the *limit* of ax is zero when the variable x is zero.

7. If
$$u = \frac{a}{x},$$
the quotient becomes greater as x diminishes in value. Hence
$$u = \text{infinity}$$
when x becomes equal to zero.

We may illustrate this in the following manner: As division is a repeated subtraction of equal quantities from a given sum, suppose we have ten dollars: we say, "two would go into ten five times;" or, we could go five times and take two dollars at each time. Now, how often could we go and take *nothing?* Evidently an infinite number of times.

The truth of the proposition may be proved thus:

Divide $\frac{1}{1-x}$; we shall have

$$\frac{1}{1-x} = 1+x+x^2+x^3+\text{etc., to infinity};$$

and making $x=1$,

$$\frac{1}{0} = 1+1+1+1+\text{etc., to infinity},$$

which is an infinite quantity.

$$\therefore \frac{a}{0} = \infty.$$

8. Since the value of a fraction diminishes as the denominator increases, it follows that, as the denominator approaches in value toward infinity, the value of the fraction approaches zero; hence the limit of

$$\frac{a}{x}$$

is *zero* when x is equal to *infinity*.

9. As the denominator of a fraction becomes nearer and nearer in value to the numerator, the fraction itself approximates in value to unity; and when the denominator becomes equal to the numerator, the value is unity.

That is, $$\frac{a^m}{a^n} = a^\circ = 1,$$

when $n=m$.

10. By the term *limit*, therefore, we mean *that value toward which a variable quantity is approaching, but which it never reaches so long as the number of terms is finite.*

We shall illustrate the foregoing by a few examples.

Ex. 1. Find the limit of the value of the fraction

$$\frac{2x+5}{3x+6};$$

first, when $x=0$, and then when $x=\infty$.

When $x=0$, $2x$ and $3x$ both become zero, and the fraction is $\frac{5}{6}$.

Again, when $x=\infty$, divide both numerator and denominator by x, and we have

$$\frac{2+\dfrac{5}{x}}{3+\dfrac{6}{x}}.$$

E 2

Now, when $x = \infty$, $\dfrac{5}{x}$ and $\dfrac{6}{x}$ each become zero, and the fraction is $\frac{2}{3}$.

Ex. 2. What is the limit to which the ratio of $2xh + h^2$ to h approaches, as h diminishes and ultimately becomes equal to zero? That is, to find the value of
$$\frac{2xh + h^2}{h}$$
when $h = 0$.

By dividing both numerator and denominator by h, we have
$$\frac{2x + h}{1};$$
and when $h = 0$, the ratio is $2x : 1$.

Ex. 3. What is the limit of
$$\frac{3x - 7}{7x - 8}$$
when $x = 0$ and $x = $ infinity?

11. *As the arc of a circle diminishes, and ultimately becomes equal to zero, the ratio of the sine, tangent, or chord of the arc to the arc itself is a ratio of equality.*

Let B'B be the side of a regular polygon inscribed in a circle, and D'D the corresponding side of the regular polygon circumscribed about the circle. Assuming the radius of the circle equal to unity, we shall have

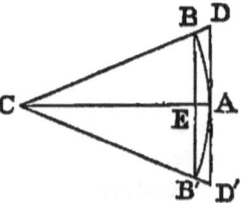

BE = sin. BCA; ∴ B'B = 2 sin. BCA.
DA = tan. BCA; ∴ D'D = 2 tan. BCA.

Now if $n = $ the number of sides of the polygon, the angle BCA will be equal to $\dfrac{\pi}{n}$, and

$$B'B = 2 \sin. \frac{\pi}{n},$$

$$D'D = 2 \tan. \frac{\pi}{n};$$

therefore the perimeter of the inscribed polygon will be
$$2n \sin. \frac{\pi}{n},$$

and the perimeter of the circumscribing polygon will be

$$2n \tan. \frac{\pi}{n}.$$

Now if we call p the perimeter of the inscribed and P the perimeter of the circumscribed polygon, we shall have

$$\frac{p}{P} = \frac{\sin. \frac{\pi}{n}}{\tan. \frac{\pi}{n}} = \cos. \frac{\pi}{n};$$

and as n, the number of sides, increases indefinitely, the angle subtended by one of these sides may become less than any assignable angle whatever, and therefore the $\cos. \frac{\pi}{n}$ becomes ultimately cos. $0 = 1$.

$$\therefore \frac{p}{P} = 1, \text{ or } p = P.$$

Hence an nth part of the perimeter of the polygon is equal to the nth part of the circumference of the circle. That is,

$$2 \sin. \frac{\pi}{n} = \frac{2\pi}{n} = 2 \tan. \frac{\pi}{n},$$

or $\quad \sin. \frac{\pi}{n} = \frac{\pi}{n}, \tan. \frac{\pi}{n} = \frac{\pi}{n};$ also chd. $\frac{\pi}{n} = \frac{\pi}{n}.$

Put $\qquad \frac{\pi}{n} = \theta.$

Then $\qquad \frac{\sin. \theta}{\theta} = 1, \frac{\tan. \theta}{\theta} = 1, \frac{\text{chd. } \theta}{\theta} = 1.$

12. From what precedes, it will be understood that by the term *limit*, or *limiting value*, we mean a fixed quantity to which a variable quantity continually approaches by making certain suppositions, and which may be made to approach so near to it that the difference shall be less than any assignable quantity. These suppositions are, generally, that one of the quantities becomes indefinitely great or indefinitely small; that is, becomes infinite or nothing, or that some of the variables become equal to constant quantities.

13. When we speak of the *limiting* value of a fraction, we do not mean *the limit of the numerator* divided by

the limit of the denominator, but *the limit of the quotient* resulting from actually dividing the numerator by the denominator. Thus, if we wish to find the limit of the fraction $\dfrac{a^2-x^2}{a-x}$, when the variable x approaches the value of a, and ultimately becomes equal to it, we would not say
$$\frac{a^2-a^2}{a-a}=\frac{0}{0},$$
for this is not the limit we are seeking. But
$$\frac{a^2-x^2}{a-x}=\frac{(a+x)(a-x)}{a-x}=a+x;$$
and when $x=a$, the value or limit is $2a$.

14. If from the point A two secants, AB, AF, be drawn, cutting the circle, then (Geom., B. II., Theor. XXI.)
$$AB \times AD = AF \times AE.$$
Put $\qquad AE=x,$
and $\qquad EF=h;$
then $\qquad AF=x+h,$
and $\qquad AB \times AD = x(x+h).$

Now let AF move to the right till it takes the position of AG, a tangent to the circle; then EF, or h, becomes zero, and $AB \times AD = x^2$; or $AB : x :: x : AD$.

Hence, by this application of the theory of limits we demonstrate the property, that *if from a point without a circle a tangent and secant be drawn, the tangent will be a mean proportional between the secant and its external segment* (Geom., B. II., Theor. XXI., Cor. 1).

EXAMPLES.

Ex. 4. What is the limiting value of $\dfrac{a^3-x^3}{a-x}$ when $x=a$?

Ans. $3a^2$.

Ex. 5. What is the limiting value of $\dfrac{x+1}{2x+1}$ when $x=\infty$?

Ans. $\tfrac{1}{2}$.

Ex. 6. What is the limiting value of $\dfrac{hx+h^2}{h}$ when $h=0$?

Ans. x.

Ex. 7. What is the limiting value of $\dfrac{3x^2h+3xh^2+h^3}{h}$ when $h=0$? Ans. $3x^2$.

Ex. 8. What is the limit of the value of $\dfrac{ax}{a+x}$ when x becomes infinite?

Ex. 9. What is the limiting value of $\dfrac{B}{A}\sqrt{2Ax+x^2}$ when x becomes infinite?

$$\dfrac{B}{A}\sqrt{2Ax+x^2}=\dfrac{B}{A}x\sqrt{\dfrac{2A}{x}+1}.$$

Now, when x becomes infinite, the quantity $\dfrac{2A}{x}=0$; therefore the limiting value $=\dfrac{B}{A}\cdot x$.

15. If a variable quantity increase uniformly, then other quantities depending on this and constant quantities may either increase uniformly, or according to any variable law whatever.

1°. Thus, if a variable quantity, x, increase uniformly, then $2x$, $3x$, or any given number of times x, will increase uniformly.

2°. Let x increase uniformly by *one*, then its successive values will be $x+1$, $x+2$, $x+3$, etc. Or, if we take ax and add a constant quantity to it, the values will go on increasing *uniformly*.

Thus, $ax+b$, $ax+2b$, $ax+3b$, etc.

16. But it most frequently happens that the quantities we have to consider *do not increase uniformly*, while the independent variable increases at a uniform rate.

For, let x increase uniformly by *one*, so as to become $x+1$, $x+2$, $x+3$, etc., then the values of the squares of these quantities do not increase uniformly.

By squaring these quantities, the successive values will be

$$(x+1)^2=x^2+2x+1,$$
$$(x+2)^2=x^2+4x+4,$$
$$(x+3)^2=x^2+6x+9,$$
etc., etc.,

in which the successive differences are

$2x+1,$
$2x+3,$
$2x+5,$
etc.,
which go on constantly increasing.

17. *When quantities are so connected that the value of each is dependent upon that of the others, each is said to be a function of the others* (Alg., Art. 169).

Thus the *area* of a square depending on the length of its side is said to be a *function of the side*.

The algebraic expressions

$$ax^2, \quad \sqrt{a^2+x^2}, \quad \frac{a+bx^3}{ax}, \text{ etc.,}$$

depending for their value on that which we assign to x, are all functions of x.

By the term *function of* x, then, we mean any algebraic expression into which x enters in combination with constant quantities.

If the expression be put equal to a single letter, as

$$u = \sqrt{a^2+x^2},$$

x is called the *independent variable*, and u, or its equal, the *dependent variable*.

18. The relation between a function and its variable is generally expressed thus:

$$u=f(x), \quad u=f'(x), \quad u=\phi(x),$$

where each expression is read, "u is equal to a function of x."

19. *Functions* are either *algebraic* or *transcendental;* these, again, are either *explicit* or *implicit, increasing* or *decreasing*.

20. An *algebraic* function is one in which the relation between it and its variable is expressed by the *sum* or *difference, product* or *quotient, roots* or *powers* of the variables, the roots or powers being *constant quantities*. Thus,

$$u = ax^2 + bx + c,$$
$$u = (bx^3 + cx^2 - ax)^{\frac{1}{2}},$$

are *algebraic* functions.

21. A *transcendental* function is one in which the relation between the function and its variable can not be expressed in the usual algebraic terms. Thus,

DEFINITIONS OF FUNCTIONS. 111

$$u = \sin x,$$
$$u = \tan x,$$
$$u = \sec x,$$

which are called *circular* functions;
or
$$u = \log x,$$
which is a *logarithmic* function;
or
$$u = a^x,$$
which is called an *exponential* function.

22. An *explicit* function is one in which the value of the function is directly expressed in terms of the variable, as
$$u = mx + n.$$

23. An *implicit* function is one in which the value of the function is not directly expressed in terms of its variable, as
$$au^2 + 2ux = bx^2.$$

24. An *increasing* function is one which increases as its variable increases, and decreases as its variable decreases, as
$$u = ax^2 + bx,$$
$$u = \sin x.$$

25. A *decreasing* function is one which decreases as its variable increases, and increases as its variable decreases, as
$$u = \frac{a}{x},$$
$$u = \cos x.$$

26. The mutual dependence of one variable on another is generally expressed under the form
$$f(x, y) = 0,$$
in which x and y are functions of each other.

27. Every equation containing two or more quantities expresses a *relation* among those quantities. If, therefore, we make any one of these quantities *dependent* on the others, the quantity thus made dependent may be called a *function* of the others. Take, for instance, the equation of the circle when the origin of co-ordinates is at the centre,
$$y^2 + x^2 = R^2.$$
As R is constant, we may make y to depend on x, or we may make x dependent on y. Thus,
$$y = \pm \sqrt{R^2 - x^2},$$
or
$$x = \pm \sqrt{R^2 - y^2}.$$

It is evident that x and y both vary from $+R$ to $-R$, and that for every value we assign to the one between these limits, there will be two values of the other numerically equal, with contrary signs.

28. When one or more quantities vary, any function of these quantities varies also, the quantity and manner of this variation being dependent on the nature of the function. *This kind of dependent variation is the first idea the student has to master*, and he will be greatly facilitated in his progress if he can acquire that knowledge from analytical relations alone.

Let us take
$$y = 2x - 5;$$
then, supposing x to vary through all magnitudes from $-\infty$ to $+\infty$, y will also vary through all magnitudes from $-\infty$ to $+\infty$.

By taking values of x that have very small differences, the *mind* can easily acquire the idea of the variation of x and y when passing through magnitudes which have insensible differences.

29. *A variable quantity, in changing its sign, must pass through zero or infinity.* Thus, in the equation
$$y = 2x - 5,$$
while x varies by insensible differences from 1 to 3, y varies from -3 to $+1$, and must, during this variation, pass through zero. This takes place when $x = 2.5$.

30. The converse of this proposition, however, does not follow. That is, *a variable function, in passing through zero or infinity, does not necessarily change its sign*. For, if we have the function
$$y = (2x - 5)^2,$$
it could never change its sign; but while x varies from $-\infty$ to $+\infty$, y will decrease from $+\infty$ to 0, then increase from 0 to $+\infty$.

The value of x that makes $y = 0$ is, as above, equal 2.5.

31. The whole theory of the variation of variables and their functions constitutes the science called the DIFFERENTIAL CALCULUS.

32. One great object of the Differential Calculus, and an object which is continually occurring, is to find *the limit of the ratio of the increment of the function to that of its variable.*

33. This limiting ratio of the increment of the function

DIFFERENTIATION OF FUNCTIONS. 113

to that of its variable is called the *differential coefficient*, because it is the coefficient of the differential of the independent variable.

34. Thus, if $f(x)$ denotes any function of x, and we give to the independent variable x any arbitrary increment, h, we shall have
$$f(x+h)$$
as the same function of $(x+h)$, and the increment of the function is evidently
$$f(x+h) - f(x).$$
Now the ratio of the increment of the function to that of the variable is
$$\frac{f(x+h) - f(x)}{h}. \qquad (1)$$
Then, if we let h diminish without limit, this ratio is the differential coefficient of $f(x)$ with respect to its variable x.

35. The symbol used to represent the differential coefficient of $f(x)$ is
$$\frac{d \cdot f(x)}{dx};$$
or, if we put $u = f(x)$, then the symbol is
$$\frac{du}{dx}.$$

The numerator of the fraction (1) represents, as we have said before, the increment of the function; the denominator is the increment of the variable. Hence the fraction expresses the ratio of the increment of the function to that of the variable; and when we pass to the limit by making $h=0$, the numerator becomes the differential of the function, which we represent by $df(x)$, and the denominator becomes the differential of the variable, which we represent by dx.

In both cases the letter d is a mere symbol, denoting "the differential of."

36. Let $\qquad u = f(x) = ax.$ \qquad (1)

Now, if we give to x an increment, h, u will change its value, and its new value may be expressed by u'. We shall then have
$$u' = f(x+h) = a(x+h). \qquad (2)$$
And, by subtracting (1) from (2),

$$u'-u = f(x+h) - f(x) = ah.$$
Or, by dividing by h,
$$\frac{u'-u}{h} = \frac{f(x+h)-f(x)}{h} = a.$$

Here the member $\frac{f(x+h)-f(x)}{h}$, which we have also represented by $\frac{u'-u}{h}$, expresses the ratio of the increment of the function to that of its variable. Hence, if we pass to the limit by making $h=0$, we shall have
$$\frac{du}{dx} \text{ or } \frac{df(x)}{dx} = a. \tag{3}$$

The first of these forms, being more easily printed, will frequently be used instead of the second.

Multiplying both members by dx, we shall have
$$du = a\,dx. \tag{4}$$

37. In (3) the first member, $\frac{du}{dx}$ or $\frac{df(x)}{dx}$, is the symbol of the differential coefficient of u with respect to x, and the second member, a, is the differential coefficient itself.

38. Again: suppose $f(x) = ax^2$, and let x become $x+h$; then
$$f(x+h) = a(x+h)^2,$$
$$= ax^2 + 2axh + ah^2;$$
and
$$f(x+h) - f(x) = 2axh + ah^2,$$
or
$$\frac{f(x+h)-f(x)}{h} = 2ax + ah.$$

Here the first member expresses the ratio of the increment of the function to that of its variable. Therefore, by passing to the limit, we have
$$\frac{df(x)}{dx} = 2ax,$$
which is the differential coefficient of ax^2.

Hence *the differential of a constant multiplied by a variable function is equal to the constant into the differential of the function.*

39. Let $f(x) = \frac{a}{x}$, and give to x an increment, h; then

DIFFERENTIATION OF FUNCTIONS. 115

$$f(x+h) = \frac{a}{x+h},$$
$$f(x+h) - f(x) = -\frac{ah}{x(x+h)},$$
or
$$\frac{f(x+h) - f(x)}{h} = -\frac{a}{x(x+h)},$$
and passing to the limit
$$\frac{df(x)}{dx} = -\frac{a}{x^2},$$
which is the differential coefficient of $\frac{a}{x}$.

40. *The differential of a constant connected by the sign plus or minus to a variable is always zero.*

Suppose $u = x^2 \pm a$,
and let x become $x+h$, then u will become u';
$$\therefore u' = (x+h)^2 \pm a = x^2 + 2xh + h^2 \pm a,$$
and
$$u' - u = 2xh + h^2,$$
or
$$\frac{u' - u}{h} = 2x + h;$$
$$\therefore \frac{du}{dx} = 2x.$$

41. From this we perceive *that the differential of a variable function and the differential of the same function increased or diminished by a constant quantity are both equal.*

42. *The differential of the sum or difference of any number of functions depending on the same variable is equal to the sum or difference of their differentials taken separately.*

Suppose $u = f(x) + f'(x) - f''(x)$,
where $f(x)$, $f'(x)$, and $f''(x)$ are different functions of x, and give to x an increment, h; then u will become u'.
$$\therefore u' = f(x+h) + f'(x+h) - f''(x+h),$$
$$u' - u = f(x+h) - f(x) + f'(x+h) - f'(x) - f''(x+h) + f''(x).$$

And, by dividing by h,
$$\frac{u' - u}{h} = \frac{f(x+h) - f(x)}{h} + \frac{f'(x+h) - f'(x)}{h} - \frac{f''(x+h) - f''(x)}{h}.$$

The first member of this equation expresses the ratio of the increment of the *whole* function to that of its variable, and the terms in the second member express the ratio of the increment of *each* function, respectively, to that of its variable. Therefore, by passing to the limit, we have

$$\frac{du}{dx} = \frac{df(x)}{dx} + \frac{df'(x)}{dx} - \frac{df''(x)}{dx},$$

or $\quad du = df(x) + df'(x) - df''(x).$

43. *The differential of the product of two functions depending on the same variable is equal to the first into the differential of the second plus the second into the differential of the first.*

Let $\quad u = f(x) f'(x),\quad\quad\quad (1)$
each being different functions of x, and give to x an increment, h; then

$$u' = f(x+h) f'(x+h), \quad\quad (2)$$
and $\quad u' - u = f(x+h) f'(x+h) - f(x) f'(x). \quad (3)$

Subtract and add the quantity $f(x) f'(x+h)$ from and to the second member of (3), we shall have

$$u' - u = f(x+h) f'(x+h) - f(x) f'(x) - f(x) f'(x+h) + f(x) f'(x+h);$$

or, by factoring, $\quad u' - u =$
$\quad f'(x+h)\{f(x+h) - f(x)\} + f(x)\{f'(x+h) - f'(x)\}.$
Now divide by h,

$$\frac{u'-u}{h} = f'(x+h) \cdot \frac{f(x+h)-f(x)}{h} + f(x) \cdot \frac{f'(x+h)-f'(x)}{h}$$

Passing to the limit, we have

$$\frac{du}{dx} = f'(x) \cdot \frac{df(x)}{dx} + f(x) \cdot \frac{df'(x)}{dx},$$

or $\quad du = f'(x) \cdot df(x) + f(x) \cdot df'(x).$

If we divide this last equation by the first, member by member, we shall have

$$\frac{du}{u} = \frac{df(x)}{f(x)} + \frac{df'(x)}{f'x}.$$

44. Hence *the differential of the product of two functions divided by their product is equal to the sum of the quotients which arises by dividing the differential of each function by its function.*

This holds good for the product of any number of functions.

DIFFERENTIATION OF FUNCTIONS.   117

Let $u = wyz$,
where w, y, z are each functions of x.
Put $wy = v$;
then $u = vz$.
Then, by the preceding,
$$\frac{du}{u} = \frac{dv}{v} + \frac{dz}{z},$$
$$\frac{dv}{v} = \frac{dw}{w} + \frac{dy}{y}.$$
Substitute for $\frac{dv}{v}$ its value, we have
$$\frac{du}{u} = \frac{dw}{w} + \frac{dy}{y} + \frac{dz}{z}.$$
Clearing this of fractions, and taking wyz for the value of u, we shall have
$$du = d.(wyz) = yzdw + wzdy + wydz.$$

45. Hence *the differential of the product of any number of functions is equal to the sum of the products of the differential of each by the product of all the others.*

46. *The differential of a fraction is equal to the denominator into the differential of the numerator minus the numerator into the differential of the denominator divided by the square of the denominator.*

Suppose $u = \frac{f(x)}{f'(x)}$,

and let x become $x + h$.

Then $u' = \frac{f(x+h)}{f'(x+h)}$,

and $u' - u = \frac{f(x+h)}{f'(x+h)} - \frac{f(x)}{f'(x)}$,

$$= \frac{f'(x)f(x+h) - f(x)f'(x+h)}{f'(x)f'(x+h)}.$$

If we add to and subtract from the numerator of this fraction the quantity $f(x)f'(x)$, we shall have

$$u' - u = \frac{f(x+h)f'(x) - f(x)f'(x+h) + f(x)f'(x) - f(x)f'(x)}{f'(x)f'(x+h)}.$$

Put the numerator into factors,

$$u' - u = \frac{f'(x)\{f(x+h) - f(x)\} - f(x)\{f'(x+h) - f'(x)\}}{f'(x)f'(x+h)}.$$

Divide both members by h,

$$\frac{u'-u}{h} = \frac{f'(x) \cdot \frac{f(x+h)-f(x)}{h} - f(x) \cdot \frac{f'(x+h)-f'(x)}{h}}{f'(x)f'(x+h)}.$$

Here the first member expresses the ratio of the increment of the function to that of the variable, and, passing to the limit,

$$\frac{f(x+h)-f(x)}{h} \text{ becomes } \frac{df(x)}{dx},$$

and $\dfrac{f'(x+h)-f'(x)}{h}$ becomes $\dfrac{df'(x)}{dx}$,

and the denominator of the fraction, which is $f'(x)f'(x+h)$, becomes $(f'x)^2$;

$$\therefore \frac{du}{dx} = \frac{f'(x)\frac{df(x)}{dx} + f(x)\frac{df'(x)}{dx}}{(f'x)^2},$$

or
$$du = \frac{f'(x)df(x) - f(x)df'(x)}{\{f'(x)\}^2}.$$

47. Turning to Art. 39, we see that the differential coefficient of $\dfrac{a}{x}$ is $-\dfrac{a}{x^2}$. Now $\dfrac{a}{x}$ is a *decreasing* function, and its differential coefficient is negative. We may easily show that *the differential coefficient of every decreasing function is always negative*.

Let $\quad u = f(x)$,
$f(x)$ being a decreasing function. Then let x become $x+h$. Then

$$u' = f(x+h),$$

and $\quad u' - u = f(x+h) - f(x),$

or $\quad \dfrac{u'-u}{h} = \dfrac{f(x+h)-f(x)}{h}.$

Now as $u > u'$, the second member must be essentially negative;

$$\therefore \frac{du}{dx} = -\frac{df(x)}{dx}.$$

48. *The differential of any constant power of a variable is equal to the exponent of the power into the variable raised to a power less by unity than the primitive exponent, and that product by the differential of the variable.*

DIFFERENTIATION OF FUNCTIONS. 119

Let $u = x^n$,
and give to x an increment, h; we shall then have
$$u' = (x+h)^n,$$
and $u' - u = (x+h)^n - x^n.$

Now the difference between the same powers of any two quantities is always divisible by the difference of those quantities (Alg., Art. 144).

$$\therefore u' - u = \{x + h - x\}\{(x+h)^{n-1} + (x+h)^{n-2}x + \ldots x^{n-1}\};$$

or, dividing by h,

$$\frac{u' - u}{h} = \{(x+h)^{n-1} + (x+h)^{n-2}x + \ldots x^{n-1}\},$$

where the second member contains n terms.

Hence, passing to the limit,
$$\frac{du}{dx} = nx^{n-1};$$
$$\therefore du = nx^{n-1}dx.$$

49. *The differential of any constant power of a function is equal to the exponent multiplied by the function raised to a power less by unity than the primitive exponent, and then by the differential of the function.*

Let $u = (f(x))^n$,
where $f(x)$ denotes any function of x, and n is a positive whole number. Let x become $x+h$. Then
$$u' = (f(x+h))^n,$$
and $u' - u = (f(x+h))^n - (f(x))^n.$

As in the preceding article, separate the second member into two factors (Alg., Art. 144).

$$\therefore u' - u =$$
$$\{f(x+h) - f(x)\}\{f(x+h)^{n-1} + f(x)f(x+h)^{n-2} + \ldots f(x)^{n-1}\}.$$

Dividing by h, $\quad \dfrac{u'-u}{h} =$

$$\frac{f(x+h) - f(x)}{h}\{f(x+h)^{n-1} + f(x)f(x+h)^{n-2} + \ldots f(x)^{n-1}\}.$$

The first member expresses the ratio of the increment of the *power of the function* to the increment of the variable, and therefore, in passing to the limit, becomes the differential coefficient of the power of the function.

The expression $\dfrac{f(x+h) - f(x)}{h}$ expresses the ratio of the increment of the *first power of the function* to that of the variable, and hence becomes the differential coefficient of the function. And as there are n terms in the second factor, we shall have, by making $h=0$, $n(f(x))^{n-1}$ as the sum of those terms.

$$\therefore \frac{du}{dx} = n \cdot (f(x))^{n-1} \cdot \frac{df(x)}{dx},$$

or $\qquad du = n \cdot (f(x))^{n-1} d \cdot f(x).$

We have thus proved the rule for the differentiation of a positive integral power of a function.

Again: let $\quad u = (f(x))^{-n},$
where n is integral; then

$$u = \frac{1}{(f(x))^n},$$

or $\qquad u(f(x))^n = 1.$

The first member is the product of two variable functions, and (by Art. 43) we have

$ud(f(x))^n + (f(x))^n du = 0$, the dif. of 1 being $=0$.

But the differential of $(f(x))^n = n(f(x))^{n-1} df(x)$;

$$\therefore un(f(x))^{n-1} df(x) + (f(x))^n du = 0;$$

$$\therefore du = -\frac{un(f(x))^{n-1} df(x)}{(f(x))^n}$$

$$= -n(f(x))^{-n-1} df(x),$$

by substituting for u its value $(f(x))^{-n}$.

Again: let $\quad u = (f(x))^{\frac{m}{n}};$
then $\qquad u^n = (f(x))^m.$

And by Art. 49,

$$nu^{n-1} du = m(f(x))^{m-1} df(x),$$

$$du = \frac{m}{n} \{f(x)\}^{m-1} df(x) \div u^{n-1};$$

$$\therefore du = \frac{m}{n} \cdot \{f(x)\}^{\frac{m}{n}-1} d \cdot f(x).$$

If $\frac{m}{n} = \frac{1}{2}$, then the differential of $(f(x))^{\frac{1}{2}}$ will be

$$\tfrac{1}{2} \{f(x)\}^{-\frac{1}{2}} df(x), \text{ or } \frac{df(x)}{2\sqrt{f(x)}}.$$

That is, *the differential of a radical function of the second degree is equal to the differential of the function under the radical sign divided by twice the radical function.*

DIFFERENTIATION OF FUNCTIONS.   121

50. *The differential coefficient which is obtained by regarding u as a function of x, is equal to the reciprocal of the differential coefficient which is obtained by regarding x as a function of u.*

Let
$$u=f(x), \qquad (1)$$
and deduce from this equation the value of x in terms of u, and let it be represented by the equation
$$x=f'(u). \qquad (2)$$
In (1) let x become $x+h$, then u will become $u+k$, and
$$u+k=f(x+h). \qquad (3)$$
Now in (2) let the value of x be $x+h$, then
$$x+h=f'(u+k). \qquad (4)$$
Subtract (1) from (3),
$$k=f(x+h)-f(x). \qquad (5)$$
Divide both members of (5) by h, then
$$\frac{k}{h}=\frac{f(x+h)-f(x)}{h}. \qquad (6)$$
Subtract (2) from (4), then
$$h=f'(u+k)-f'(u). \qquad (7)$$
Divide both members of (7) by k, then
$$\frac{h}{k}=\frac{f'(u+k)-f'(u)}{k}. \qquad (8)$$

In (6) and (8) the same symbols have the same values. Hence, multiplying (6) and (8) together,
$$\frac{k \cdot h}{h \cdot k}=\frac{f(x+h)-f(x)}{h} \cdot \frac{f'(u+k)-f'(u)}{k},$$
or
$$1=\frac{f(x+h)-f(x)}{h} \cdot \frac{f'(u+k)-f'(u)}{k}.$$

Each factor in the second member expresses the ratio of the increment of the function to that of its variable; therefore, passing to the limit, we have
$$1=\frac{df(x)}{dx} \cdot \frac{df'(u)}{du}.$$
But $f(x)=u$, and $f'(u)=x$; hence
$$1=\frac{du}{dx} \cdot \frac{dx}{du},$$
or
$$\frac{du}{dx}=\frac{1}{\frac{dx}{du}}.$$

F

DIFFERENTIAL COEFFICIENT OF A FUNCTION OF A FUNCTION.

51. *If three quantities are mutually dependent on each other, the differential coefficient of the first regarded as a function of the third, will be equal to the differential coefficient of the first regarded as a function of the second multiplied by the differential coefficient of the second regarded as a function of the third.*

For, let
$$u = f(y), \quad (1)$$
and
$$y = f'(x). \quad (2)$$

If we give to x an increment, h, we shall have
$$y' - y = f'(x+h) - f'(x). \quad (3)$$

Put this increment of y equal to k, and substitute it in (Eq. 1). Then
$$u' = f(y+k),$$
and
$$u' - u = f(y+k) - f(y). \quad (4)$$

Divide (3) by h and (4) by k,
$$\frac{y'-y}{h} = \frac{f'(x+h) - f'(x)}{h}, \quad (5)$$

$$\frac{u'-u}{k} = \frac{f(y+k) - f(y)}{k}. \quad (6)$$

Multiply (5) and (6) together, member by member,
$$\frac{u'-u}{k} \cdot \frac{y'-y}{h} = \frac{f(y+k) - f(y)}{k} \cdot \frac{f'(x+h) - f'(x)}{h}.$$

But $y'-y$ was made equal to k; therefore
$$\frac{u'-u}{h} = \frac{f(y+k) - f(y)}{k} \times \frac{f'(x+h) - f'(x)}{h}.$$

Each one of these expressions represents the ratio of the increment of its function to that of its variable respectively. Hence, passing to the limit, we have
$$\frac{du}{dx} = \frac{df(y)}{dy} \cdot \frac{df'(x)}{dx}.$$

But $f(y) = u$, and $f'(x) = y$,
$$\therefore \frac{du}{dx} = \frac{du}{dy} \cdot \frac{dy}{dx}.$$

EXAMPLES.

Ex. 10. If x increase uniformly at the rate of 2, at what rate does ax^2 increase when $a = 4$ and $x = 10$?

DIFFERENTIATION OF FUNCTIONS. 123

Let $u = ax^2$;
then $du = 2ax\,dx$.
Here $a = 4$, $x = 10$, and $dx = 2$;
$\therefore du = 160$.

Ex. 11. If the side of a square increase uniformly at the rate of 3 inches a second, at what rate is the area increasing when the side becomes 10 inches?
Let $u = x^2 =$ the area;
then $du = 2x\,dx$.
Here $x = 10$, and $dx = 3$;
$\therefore du = 60$.

Hence the area is increasing at the rate of 60 square inches a second.

Ex. 12. If the linear dimension of a cube increase uniformly at the rate of one tenth of an inch a second, at what rate does its volume increase when the edge becomes 20 inches?
Let $u = x^3 =$ volume of the cube;
then $du = 3x^2\,dx$.
Here $x = 20$ and $dx = \frac{1}{10}$;
$\therefore du = 3 \times 400 \times \frac{1}{10} = 120$ cubic inches.

Differentiate the following functions:

Ex. 13. $u = a^2 + 2ax^3 - x$.
Ex. 14. $u = (a + x)^2$.
Ex. 15. $u = a^2 - x^2$.
Ex. 16. $u = ax^{\frac{1}{2}} + \frac{1}{2}y^2 - \frac{x}{3}$.
Ex. 17. $u = 4x^4 - 3x^3 + 5x^2 - 6x$.
Ex. 18. $u = 3x^4 + bx^3 + cx^2$.
Ex. 19. $u = 7x^5 + 9x^4 - 5x^3 + 3x^2 - 4x + c$.
Ex. 20. $u = x^2 - 4x + 5\sqrt{x} - 6x^{\frac{1}{3}}$.
Ex. 21. $u = 5x^2 - 3y^2 + 4x^3 - y^4 + c$.
Ex. 22. $u = (ax + x^2)^2$.
    Ans. $du = 2(ax + x^2)(a + 2x)dx$.
Ex. 23. $u = (ax + x^2)^3$.
    Ans. $du = 3(ax + x^2)^2(a + 2x)dx$.
Ex. 24. $u = (ax^2 + x^3)^2$.
    Ans. $du = 2x^3(a + x)(2a + 3x)dx$.
Ex. 25. $u = (ax^2 + x^3)^3$.
    Ans. $du = 3x^5(a + x)^2(2a + 3x)dx$.
Ex. 26. $u = (1 + x)^2$.

Ex. 27. $u = \left(\dfrac{1+x}{a}\right)^2$.

$$Ans. \ 2 \cdot \dfrac{(1+x)\,dx}{a}.$$

Ex. 28. $u = 3x^{\frac{1}{3}} + \frac{3}{4}x^{\frac{4}{3}} - \dfrac{3ax^{\frac{1}{3}}}{b} + \frac{3}{2}x^{\frac{4}{7}}$.

Ex. 29. If the *area* of a square increase uniformly at the rate of $\frac{1}{10}$th of a square inch a second, at what rate is the *side* increasing when the area is 100 square inches?

Let $x =$ the side, and $u =$ the area;
then $u = f(x) = x^2$,
and therefore $x = f'(u) = \sqrt{u}$,

$$dx = \dfrac{du}{2\sqrt{u}} = \dfrac{du}{2x},$$

or $u = x^2$; $\therefore du = 2x\,dx$; $\therefore dx = \dfrac{du}{2x}$.

Here $du = \frac{1}{10}$, and $x = \sqrt{u} = \sqrt{100} = 10$, and $2x = 20$;
$\therefore dx = .005$ of an inch.

Ex. 30. If the *volume* of a cube increase uniformly at the rate of a cubic inch a second, at what rate is the length of a *side* increasing when the solid becomes a cubic foot? $Ans.$ $\frac{1}{432}$ of an inch per second.

Differentiate the following functions:

Ex. 31. $u = (a^2 + x^2)^3$.

Ex. 32. $u = (x + y)^2$.

Ex. 33. $u = (x + ax^2)^{\frac{1}{2}}$.

Ex. 34. $u = \sqrt{1 + x}$.

Ex. 35. $u = \sqrt[3]{a + x^2}$.

Ex. 36. $u = \left(\dfrac{a}{b} + \dfrac{x}{2}\right)^{\frac{1}{2}}$.

Ex. 37. $u = \sqrt{x^2 + a\sqrt{x}}$.

Ex. 38. $u = (x^2 + y^2)^{\frac{3}{2}}$.

Ex. 39. $u = (a^5 + x^5)^{\frac{1}{2}}$.

Ex. 40. $u = (a^2 + x^2)^{-\frac{5}{9}}$.

Ex. 41. $u = x^2 y^{\frac{1}{2}} z^3$.

Ex. 42. $u=ax^2y+x\sqrt{y}-\sqrt{xy}$.
Ex. 43. $u=x^2y^3$.
Ex. 44. $u=x^{\frac{5}{3}}y^{\frac{7}{2}}z$.
Ex. 45. $u=x^m y^n z^{\frac{1}{2}}$.
Ex. 46. $u=wxyz$.

Ex. 47. If one side of a rectangle increase at the rate of an inch a second, and the other at the rate of 2 inches a second, at what rate is the *area* increasing when the first side becomes 8 inches and the other 12?

Let $x=$ the 1st side, and $y=$ the 2d;
then $u=xy$,
$du=xdy+ydx$.
Here $x=8, y=12, dx=1, dy=2$;
∴ $du=16+12=28$ square inches.

By Art. 43, the differential of a decreasing function is always negative; therefore, if one side *diminishes*, its differential must be negative.

Ex. 48. If one side of a rectangle increase at the rate of 3 inches a second, and the other *diminish* at the rate of 2 inches a second, at what rate is the area increasing or diminishing when the first side becomes 10 inches and the other 8?

Ans. 4 square inches, *increasing*.

Differentiate the following functions:

Ex. 49. $u=\dfrac{x^3}{y^2}$.

Ex. 50. $u=\dfrac{xy}{a+x}$.

Ex. 51. $u=\dfrac{1}{\sqrt{x}}$.

Ex. 52. $u=x^2y^3$.

Ex. 53. $u=x^{\frac{1}{n}}$.

Ex. 54. $u=(a+x^2)^{\frac{1}{n}}$.

Ex. 55. $u=\dfrac{ax^{\frac{1}{2}}}{b}$.

Ex. 56. $u=\dfrac{a}{bx^{\frac{1}{n}}}$.

Ex. 57. $u = \dfrac{a}{(1-x)^n}$.

Ex. 58. $u = \dfrac{x}{1-x}$.

Ex. 59. $u = \dfrac{xy^n}{x^2 - y^2}$.

Ex. 60. $u = \dfrac{(a^2 + x^n)^4}{4}$.

Ex. 61. $u = (1 + 2x^2)(1 + 4x^2)$.
Ex. 62. $u = (1+x)^4 (1+x^2)^2$.
Ex. 63. $u = (a + bx^m)^n$.
Ex. 64. $u = (1+x^m)^n (1+x^n)^m$.

Ex. 65. $u = \left(\dfrac{x}{1+x}\right)^n$.

Ex. 66. $u = \dfrac{x^2}{(a + x^2)^2}$.

Ex. 67. $u = (1+x)(1-x)^{\frac{1}{2}}$.

Ex. 68. $u = \dfrac{x}{x + \sqrt{1+x^2}}$.

Ex. 69. $u = (ax^3 + b)^2 + (x - b)(a^2 - x^2)^{\frac{1}{2}}$.

Ex. 70. $u = \dfrac{a^4}{2\sqrt{a^2 x^2 - x^4}}$.

Ex. 71. $u = \dfrac{\sqrt{x^2 + 1} - x}{\sqrt{x^2 + 1} + x}$.

Ex. 72. $u = x(a^2 + x^2)(a^2 - x^2)^{\frac{1}{2}}$.

Ex. 73. $u = \dfrac{b}{a}\sqrt{2ax + x^2}$.

Ex. 74. $u = ax^{\frac{3}{2}} - bx^{\frac{1}{2}} + c$.

$$\therefore \dfrac{du}{dx} = \tfrac{3}{2} ax^{\frac{1}{2}} - \dfrac{b}{2\sqrt{x}}.$$

Ex. 75. $u = a^2(b+x)^2 - b^2(a-x)^2$.

$$\therefore \dfrac{du}{dx} = 2(a+b)\{ab + (a-b)x\}.$$

Ex. 76. $u = \sqrt{(ax^2 + 2bx + 3c)}$.

$$\therefore \dfrac{du}{dx} = \dfrac{ax + b}{\sqrt{ax^2 + 2bx + 3c}}.$$

DIFFERENTIATION OF FUNCTIONS. 127

Ex. 77. $u = \dfrac{x}{\sqrt{1+x^2}}.$
$$\therefore \dfrac{du}{dx} = \dfrac{1}{(1+x^2)^{\frac{3}{2}}}.$$

Ex. 78. $u = \dfrac{x^2}{\sqrt{a^3+x^3}}.$
$$\therefore \dfrac{du}{dx} = \dfrac{x(4a^3+x^3)}{2(a^3+x^3)^{\frac{3}{2}}}.$$

Ex. 79. $u = ax^m - bx^{m-1} + cx^{m-2}.$

Ex. 80. $u = \dfrac{a+x}{a+x^2}.$

Ex. 81. $u = \dfrac{1}{x}.$

Ex. 82. $u = \dfrac{1}{\sqrt{a^3+x^3}}.$

Ex. 83. $u = 3xy.$

Ex. 84. $u = (a+x^2)^n.$

Ex. 85. $u = ax^{n+1}.$

Ex. 86. $u = \dfrac{a}{x^n}.$

Ex. 87. $u = (ax^n + x^m)^3.$

Ex. 88. $u = \dfrac{1}{a(a+x)^{\frac{1}{n}}}.$

Ex. 89. $u = \dfrac{x}{yz}.$

Ex. 90. $u = \dfrac{(a+x^2)^4}{8}.$

Ex. 91. $u = 3x^2 y.$

Ex. 92. $u = \dfrac{2x^4}{a^2 - x^2}.$

Ex. 93. $u = (x^3+a)(3x^2+b).$

Ex. 94. $u = (a+x)(b+x)(c+x).$

Ex. 95. $u = \dfrac{x}{x + \sqrt{1-x^2}}.$

Ex. 96. $u = \dfrac{x^3}{(1+x)^3}.$

Ex. 97. $u=\{x^2+(a+x^2)^{\frac{1}{2}}\}^{\frac{1}{2}}$.

Ex. 98. $u=\dfrac{x^3}{(1-x^2)^{\frac{3}{2}}}$.

Ex. 99. $u=\left(\dfrac{1-\sqrt{x}}{1+\sqrt{x}}\right)^{\frac{1}{2}}$.

Ex. 100. $u=\dfrac{8b\sqrt{ax^3-x^4}}{3a}$.

Ex. 101. $u=\dfrac{(x+a)^{\frac{3}{2}}}{(x-a)^{\frac{1}{2}}}$.

Ex. 102. $u=\dfrac{3(a^2+x^2)^{\frac{1}{2}}}{(a-x^{\frac{1}{2}}}$.

Ex. 103. $u=(a^2-x^2)^{\frac{1}{2}}$.

Ex. 104. $u=\dfrac{b}{a}\sqrt{2ax-x^2}$.

Ex. 105. $u=(2n-1)x^{2n+1}-(2n+1)x^{2n-1}$.

Ex. 106. $u=ax^{2n}-bx^n+c$.

Ex. 107. $u=\dfrac{a+x}{b+x}$.

Ex. 108. $u=\dfrac{x^3}{a^2}+\dfrac{a^2}{x}$.

Ex. 109. $u=\dfrac{x^m-a^m}{x^m+a^m}$.

$$\therefore \dfrac{du}{dx}=\dfrac{2ma^m x^{m-1}}{(x^m+a^m)^2}.$$

Ex. 110. $u=\dfrac{3x^2-a^2}{(a^2+x^2)^3}$.

$$\therefore \dfrac{du}{dx}=\dfrac{12x(a^2-x^2)}{(a^2+x^2)^4}.$$

Ex. 111. $u=\dfrac{a+5bx}{(a+bx)^5}$.

$$\therefore \dfrac{du}{dx}=-\dfrac{20b^2x}{(a+bx)^6}.$$

Ex. 112. $u=\dfrac{a+3bx^2}{(a+bx^2)^3}$.

DIFFERENTIATION OF FUNCTIONS. 129

$$\therefore \frac{du}{dx} = -\frac{12b^2x^3}{(a+bx^2)^4}.$$

Ex. 113. $u = ax^{\frac{3}{2}} - bx^{\frac{1}{2}} + cx^{-\frac{1}{2}} - ex^{-\frac{5}{2}}$.

Ex. 114. $u = \left\{a + bx^{\frac{3}{2}} + cx^{\frac{2}{3}}\right\}^{\frac{1}{2}}$.

Ex. 115. $u = (a+x)(a-x)^{\frac{1}{2}}$.

$$\therefore \frac{du}{dx} = \frac{a-3x}{2\sqrt{a-x}}$$

Ex. 116. $u = (a^2 + x^2)(a^2 - x^2)^{\frac{1}{2}}$.

$$\therefore \frac{du}{dx} = \frac{x(a^2 - 3x^2)}{\sqrt{a^2 - x^2}}.$$

Ex. 117. $u = (2a^2 + 3x^2)(a^2 - x^2)^{\frac{3}{2}}$.

$$\therefore \frac{du}{dx} = -15x^3(a^2 - x^2)^{\frac{1}{2}}.$$

Ex. 118. $u = (2\sqrt{a} + \sqrt{x})(\sqrt{a} + \sqrt{x})^{\frac{1}{2}}$.

$$\therefore \frac{du}{dx} = \frac{1}{2\sqrt{x}} \cdot \frac{3\sqrt{a} + 2\sqrt{x}}{(\sqrt{a} + \sqrt{x})^{\frac{1}{2}}}.$$

Ex. 119. $u = (bx - 3a)(a + bx)^{\frac{1}{3}}$.

$$\therefore \frac{du}{dx} = \frac{4b^2x}{3(a+bx)^{\frac{2}{3}}}.$$

Ex. 120. $u = (3bx - 2a)(a + bx)^{\frac{3}{2}}$.

$$\therefore \frac{du}{dx} = \tfrac{15}{2}b^2x\sqrt{a+bx}.$$

Ex. 121. $u = (5bx^2 - 2a)(a + bx^2)^{\frac{5}{2}}$.

$$\therefore \frac{du}{dx} = 35b^2x^3(a+bx^2)^{\frac{3}{2}}.$$

Ex. 122. $u = (8a^2 - 4abx + 3b^2x^2)(a+bx)^{\frac{1}{2}}$.

$$\therefore \frac{du}{dx} = \frac{15b^3x^2}{2\sqrt{a+bx}}.$$

Ex. 123. $u = \dfrac{x}{\sqrt{a - bx^2}}$.

$$\therefore \frac{du}{dx} = \frac{a}{(a-bx^2)^{\frac{3}{2}}}.$$

Ex. 124. $u = \dfrac{\sqrt{a^2-x^2}}{x}$.

$\therefore \dfrac{du}{dx} = -\dfrac{a^2}{x^2\sqrt{a^2-x^2}}$.

Ex. 125. $u = \dfrac{x^{\frac{3}{2}}}{\sqrt{a-x}}$.

$\therefore \dfrac{du}{dx} = \dfrac{(3a-2x)\sqrt{x}}{2(a-x)^{\frac{3}{2}}}$.

Ex. 126. $u = \dfrac{b+2cx}{\sqrt{a+bx+cx^2}}$.

$\therefore \dfrac{du}{dx} = \dfrac{4ac-b^2}{2(a+bx+cx^2)^{\frac{3}{2}}}$.

Ex. 127. $u = \dfrac{(x-1)(x^4+1)^{\frac{1}{2}}}{x+1}$.

$\therefore \dfrac{du}{dx} = \dfrac{2(x^5+x^4-x^3+1)}{(x+1)^2(x^4+1)^{\frac{1}{2}}}$.

Ex. 128. $u = \left(\dfrac{1+x^2}{1-x^2}\right)^{\frac{1}{2}}$.

$\therefore \dfrac{du}{dx} = \dfrac{2x}{(1-x^2)(1-x^4)^{\frac{1}{2}}}$.

Ex. 129. $u = \dfrac{\sqrt{1+x^2}+\sqrt{1-x^2}}{\sqrt{1+x^2}-\sqrt{1-x^2}}$.

$\therefore \dfrac{du}{dx} = -\dfrac{2x}{(1-\sqrt{1-x^4})(1-x^4)^{\frac{1}{2}}}$.

CHAPTER II.

OF SUCCESSIVE DIFFERENTIAL COEFFICIENTS.—TAYLOR'S THEOREM. — MACLAURIN'S THEOREM. — DEVELOPMENT OF FUNCTIONS OF TWO OR MORE VARIABLES.—TRANSCENDENTAL FUNCTIONS.—IMPLICIT FUNCTIONS.

52. THE differential coefficient derived from any function u of the variable x may contain that variable, and hence this coefficient itself may be differentiated. We shall then obtain a *second differential coefficient*. In like manner, by differentiating this second differential coefficient (if the variable still enters in it), we shall obtain a *third differential coefficient*. Thus we may continue to find the successive differential coefficients until we arrive at a coefficient in which the variable does not enter. The differential coefficient of this will evidently be equal to *zero*.

Let
$$u = ax^3; \quad (1)$$
then, by differentiating, we have
$$\frac{du}{dx} = 3ax^2. \quad (2)$$

The first member of (2) is the *symbol* of the first differential coefficient of u regarded as a function of x, and the second member is the first differential coefficient itself.

Taking (2), and differentiating it the second time, considering dx as constant, we obtain
$$d\left(\frac{du}{dx}\right) = 2.3\,ax\,dx. \quad (3)$$

But the differential of the first member is the differential of the differential of u (which we call the second differential of u), divided by the differential of x; or
$$\frac{d^2u}{dx} = 2.3\,ax\,dx;$$
$$\therefore \frac{d^2u}{dx^2} = 2.3\,ax. \quad (4)$$

Here, as in (2), the first member is the *symbol* of the second differential coefficient, and the second member *is* the second differential coefficient itself. Now, as this contains the variable, we may differentiate it also:

thus,
$$\frac{d\left(\frac{d^2u}{dx^2}\right)}{dx}=2.3a, \quad (5)$$

which, as before, can be written
$$\frac{d^3u}{dx^3}=6a. \quad (6)$$

As this third differential coefficient does not contain the variable, its differential coefficient becomes equal to zero.

Again: if $u=x^n$, we shall have
$$\frac{du}{dx}=nx^{n-1}=\text{1st dif. coef.},$$

and
$$\frac{d\left(\frac{du}{dx}\right)}{dx}=n(n-1)x^{n-2}.$$

Since the differential of x is considered constant,
$$\frac{d\left(\frac{du}{dx}\right)}{dx} \text{ becomes } \frac{d^2u}{dx^2}=n(n-1)x^{n-2}.$$

Differentiating again,
$$\frac{d\left(\frac{d^2u}{dx^2}\right)}{dx}=\frac{d^3u}{dx^3}=n(n-1)(n-2)x^{n-3}.$$

53. If n is a *positive whole number*, there will be as many differential coefficients as there are units in n, and the $(1+n)$th differential coefficient will be equal to zero.

EXAMPLES.

1. Let $\quad u=ax^3-bx^2+cx-e;$

then $\quad \dfrac{d^3u}{dx^3}=6a,$

and $\quad \dfrac{d^4u}{dx^4}=0.$

2. Let $\quad (u-b)^2=c^2-(x-a)^2,$
to find the successive differential coefficients of u regarded as a function of x.

TAYLOR'S THEOREM.

54. *Taylor's Theorem* explains the method of developing into a series any continuous function of the sum or difference of two variables which are independent of each other, according to the ascending powers of one of them.

If we have a function of the sum of two variables, x and y, the differential coefficients will be equal whether we consider x to vary and y to remain constant, or y to vary and x to remain constant. But,

If we have a function of the difference of these two variables, the differential coefficients will be equal with opposite signs.

1°. Let $u = f(x+y) = (x+y)^n$;
then, considering x variable and y constant, we shall have
$$\frac{du}{dx} = n(x+y)^{n-1}.$$
And if we suppose y to vary and x remain constant, we shall have
$$\frac{du}{dy} = n(x+y)^{n-1},$$
the same as under the first supposition.

2°. Again, let $u = f(x-y) = (x-y)^n$;
then, on the supposition that x varies and y remains constant, we have
$$\frac{du}{dx} = n(x-y)^{n-1}.$$
But if we suppose y to vary and x to be constant, we shall have
$$\frac{du}{dy} = -n(x-y)^{n-1}.$$
Here the differential coefficients are equal with opposite signs.

55. *Any continuous function of the sum or difference of two variables may be developed into a series of the following forms, viz.:*

$$f(x+y) = u + \frac{du}{dx}y + \frac{d^2u}{dx^2}\frac{y^2}{1.2} + \frac{d^3u}{dx^3}\frac{y^3}{1.2.3} +, \text{ etc. };$$

$$f(x-y) = u - \frac{du}{dx}y + \frac{d^2u}{dx^2}\frac{y^2}{1.2} - \frac{d^3u}{dx^3}\frac{y^3}{1.2.3} +, \text{ etc.},$$

in which u represents the value of the function when $y=0$.

Let u' be a function of $x+y$, which we will suppose to be developed into a series, and arranged according to the powers of y, so that we have

$$u' = f(x+y) = A_0 + A_1 y + A_2 y^2 + A_3 y^3 +, \text{ etc.}, \quad (1)$$

in which A_0, A_1, A_2, etc., are independent of y, but functions of x, and dependent on all the constants which enter into the function. We are required to find such values for A_0, A_1, etc., as shall render the development true for all possible values that may be attributed to x and y.

By differentiating (1) under the supposition that x is variable and y constant, we have

$$\frac{du'}{dx} = \frac{dA_0}{dx} + \frac{dA_1}{dx}y + \frac{dA_2}{dx}y^2 + \frac{dA_3}{dx}y^3 +, \text{ etc.}$$

And differentiating the same under the supposition that y varies and x is constant, we have

$$\frac{du'}{dy} = A_1 + 2A_2 y + 3A_3 y^2 + 4A_4 y^3 +, \text{ etc.}$$

But, by Art. 54,

$$\frac{du'}{dx} = \frac{du'}{dy}.$$

Hence we must have

$$\frac{dA_0}{dx} + \frac{dA_1}{dx}y + \frac{dA_2}{dx}y^2 + \frac{dA_3}{dx}y^3 +, \text{ etc.} =$$
$$A_1 + 2A_2 y + 3A_3 y^3 + 4A_4 y^4 +, \text{ etc.}$$

And since the coefficients are independent of y, and the equality exists whatever be the value of y, we shall have, by the Theory of Indeterminate Coefficients (Alg., Art. 144),

$$\frac{dA_0}{dx} = A_1, \qquad (2)$$

$$\frac{dA_1}{dx} = 2A_2, \qquad (3)$$

$$\frac{dA_2}{dx} = 3A_3, \qquad (4)$$

etc. = etc.

TAYLOR'S THEOREM. 135

Now, make $y=0$ in equation (1), and $f(x+y)$ will become $f(x)$, which we may denote by u. Hence
$$A_0 = u.$$
Substituting this value of A_0 in (2), we have
$$A_1 = \frac{du}{dx}.$$
Substituting this value of A_1 in (3), we have
$$A_2 = \frac{1}{1 \cdot 2} \frac{d^2u}{dx^2}.$$
In like manner,
$$A_3 = \frac{1}{1 \cdot 2 \cdot 3} \frac{d^3u}{dx^3}.$$
Finally, substituting these values of A_0, A_1, etc., in (1), we have

TAYLOR'S THEOREM:
$$u' = f(x+y) = u + \frac{du}{dx} y + \frac{d^2u}{dx^2} \frac{y^2}{1 \cdot 2} + \frac{d^3u}{dx^3} \frac{y^3}{1 \cdot 2 \cdot 3} +, \text{etc.}$$

By making $\frac{du'}{dx} = -\frac{du'}{dy}$, we would obtain the development of $f(x-y)$.

56. Taylor's Theorem may be applied to the development of the second state of any function of the form
$$u = f(x),$$
when x becomes $x+h$.

For, by substituting h for y, we have
$$u' = u + \frac{du}{dx} h + \frac{d^2u}{dx^2} \frac{h^2}{1 \cdot 2} + \frac{d^3u}{dx^3} \frac{h^3}{1 \cdot 2 \cdot 3} +, \text{etc.,}$$
or
$$u' - u = \frac{du}{dx} h + \frac{d^2u}{dx^2} \frac{h^2}{1 \cdot 2} + \frac{d^3u}{dx^3} \frac{h^3}{1 \cdot 2 \cdot 3} +, \text{etc.;}$$
$$\frac{u' - u}{h} = \frac{du}{dx} + \left(\frac{d^2u}{dx^2} \frac{1}{1 \cdot 2} + \frac{d^3u}{dx^3} \frac{h}{1 \cdot 2 \cdot 3} +, \text{etc.}\right) h.$$

It is evident that h may be made so small that the term
$$\left(\frac{d^2u}{dx^2} \frac{1}{1 \cdot 2} + \frac{d^3u}{dx^3} \frac{h}{1 \cdot 2 \cdot 3} +, \text{etc.}\right) h$$
shall be less than any assignable quantity, and therefore less than $\frac{du}{dx}$,

or
$$\frac{du}{dx} > h\left(\frac{d^2u}{dx^2}\frac{1}{1.2} + \frac{d^3u}{dx^3}\frac{h}{1.2.3} +, \text{etc.}\right).$$
Or, if we multiply both members of this inequality by h, we shall have
$$\frac{du}{dx}h > \left(\frac{d^2u}{dx^2}\frac{h^2}{1.2} + \frac{d^3u}{dx^3}\frac{h^3}{1.2.3} +, \text{etc.}\right).$$

57. Hence, *if we have a series expressed in the ascending powers of the variable, we may assign to that variable so small a value that the first term shall be greater than the sum of all the remaining terms, and the sign of the series will depend upon the sign of its first term.*

58. In the foregoing demonstration of Taylor's Theorem, we have supposed all the functions to be continuous; *but if the function to be developed be infinite for values of the variable lying between certain limits, the demonstration is not valid.*

59. A function of a variable is said to be *continuous* between certain limits of the variable when the function changes *gradually* as the variable passes from one limit to the other; that is, an indefinitely small change in the variable causes an indefinitely small change in the function. Hence a *continuous* function can never become *infinite* between the limits for which it is continuous, inasmuch as an indefinitely small change in the variable would then cause a change in the function *not indefinitely small.*

MACLAURIN'S THEOREM.

60. *If u represents a function of x which can be developed in a series of positive ascending powers of that variable, then we shall have for that development*

$$u = f(x) = U + U'x + U''\frac{x^2}{1.2} + U'''\frac{x^3}{1.2.3} +, \text{etc.,}$$

in which U, U', U'', etc., represent the values that the function of x and its successive differential coefficients respectively assume when x is made equal to *zero*.

Let $u = f(x)$,
and put $\quad u = A_0 + A_1 x + A_2 x^2 + A_3 x^3 +$, etc.; $\quad$ (1)
then $\quad \dfrac{du}{dx} = A_1 + 2A_2 x + 3A_3 x^2 + 4A_4 x^3 +$, etc., $\quad$ (2)

DEVELOPMENT OF FUNCTIONS, ETC.   137

$$\frac{d^2u}{dx^2} = 2A_2 + 2 \cdot 3A_3 x + 3 \cdot 4A_4 x^2 +, \text{etc.}, \quad (3)$$

$$\frac{d^3u}{dx^3} = 2 \cdot 3A_3 + 2 \cdot 3 \cdot 4A_4 x +, \text{etc.} \quad (4)$$

If in (1), (2), (3), (4), etc., we make $x = 0$,
then
$$u = A_0 = U,$$
$$\frac{du}{dx} = A_1 = U',$$
$$\frac{d^2u}{dx^2} = 2A_2 = U''; \therefore A_2 = \tfrac{1}{2}U''.$$
$$\frac{d^3u}{dx^3} = 2 \cdot 3A_3 = U'''; \therefore A_3 = \frac{1}{2 \cdot 3}U'''.$$

Substituting these values of A_0, A_1, etc., in (1), we have

MACLAURIN'S THEOREM.

$$u = f(x) = U + U'x + U''\frac{x^2}{1 \cdot 2} + U'''\frac{x^3}{1 \cdot 2 \cdot 3} +, \text{etc.}$$

DEVELOPMENT OF FUNCTIONS OF TWO OR MORE VARIABLES WHEN EACH RECEIVES AN INCREMENT.

61. In Art. 50 we have defined *successive differential coefficients;* that is, that the *second differential coefficient* of a function is the differential coefficient of the differential coefficient of that function. The differential coefficient of the second differential coefficient is called the *third differential coefficient* of that function, and so on. We shall now explain another notation.

If we have a function of two variables, as
$$u = f(x, y), \quad (1)$$
we may suppose one of them to remain constant, and differentiate the function with respect to the other.

Thus, if we suppose y to remain constant and x to vary, the differential coefficient will be
$$\frac{du}{dx} = f'(x, y); \quad (2)$$
and if we suppose x to remain constant and y to vary, the differential coefficient will be
$$\frac{du}{dy} = f''(x, y). \quad (3)$$

The differential coefficients in (2) and (3) are called *partial differential coefficients*. That in (2) is the partial differential coefficient of u regarded as a function of x, and that in (3) is the partial differential coefficient of u regarded as a function of y.

62. If we multiply both members of (2) by dx, and both members of (3) by dy, we shall have

$$\frac{du}{dx}dx = f'(x, y)dx, \qquad (4)$$

and

$$\frac{du}{dy}dy = f''(x, y)dy. \qquad (5)$$

The expressions $\frac{du}{dx}dx$ and $\frac{du}{dy}dy$ are called *partial differentials*; that in (4) is the partial differential of u as a function of x, and that in (5) is the partial differential of u as a function of y.

63. If we differentiate (2) under the supposition that x is variable and y constant, we shall have

$$\frac{d\left(\frac{du}{dx}\right)}{dx} = f'''(x, y),$$

or

$$\frac{d^2u}{dx^2} = f'''(x, y); \qquad (6)$$

and if we differentiate (2) under the supposition that x is constant and y remains variable, we shall have

$$\frac{d\left(\frac{du}{dx}\right)}{dy} = f^{iv}(x, y);$$

$$\therefore \frac{d^2u}{dx\,dy} = f^{iv}(x, y). \qquad (7)$$

The first member of (6) denotes that the function u has been differentiated twice, both with respect to x; and the first member of (7) denotes that the same function u has been differentiated twice, once with respect to x, and once with respect to y.

If we differentiate (7) again, regarding x as the variable, we shall have

$$\frac{d^3u}{dx^2\,dy} = f^{v}(x, y); \qquad (8)$$

DEVELOPMENT OF FUNCTIONS, ETC.   139

and, regarding y as the variable,

$$\frac{d^3u}{dxdy^2}=f^{vi}(x, y). \qquad (9)$$

In (8) the symbol of the differential coefficient denotes that the function has been differentiated three times: twice with respect to x, and once with respect to y.

In (9) the symbol of the differential coefficient denotes that the function has been differentiated three times; once with respect to x, and twice with respect to y.

And generally

$$\frac{d^{m+n}u}{dx^m dy^n}=\phi(x, y)$$

denotes that the function u has been differentiated $m+n$ times; m times with respect to x, and n times with respect to y.

64. Let $\qquad u=f(x, y)$,

and give to x an increment, h, and let y be considered constant; then, by Taylor's Theorem,

$$f(x+h, y)=u+\frac{du}{dx}\frac{h}{1}+\frac{d^2u}{dx^2}\frac{h^2}{1.2}+\frac{d^3u}{dx^3}\frac{h^3}{1.2.3}+, \text{etc.}, \quad (1)$$

where u, $\frac{du}{dx}$, $\frac{d^2u}{dx^2}$, etc., are functions of x and y, and dependent on the constants that enter the $f(x, y)$.

Each of these expressions must be developed by the same theorem, by giving to y an increment, k. Then

u will become $u+\frac{du}{dy}k+\frac{d^2u}{dy^2}\frac{k^2}{1.2}+\frac{d^3u}{dy^3}\frac{k^3}{1.2.3}+$, etc.,

and $\frac{du}{dx}$ will become $\frac{du}{dx}+\frac{d^2u}{dxdy}k+\frac{d^3u}{dxdy^2}\frac{k^2}{1.2}+$, etc.,

and $\frac{d^2u}{dx^2}$ will become $\frac{d^2u}{dx^2}+\frac{d^3u}{dx^2dy}k+\frac{d^4u}{dx^2dy^2}\frac{k^2}{1.2}+$, etc.,

and $\frac{d^3u}{dx^3}$ will become $\frac{d^3u}{dx^3}+\frac{d^4u}{dx^3dy}k+\frac{d^5u}{dx^3dy^2}\frac{k^2}{1.2}+$, etc.

Substituting these values of u, $\frac{du}{dx}$, etc., in equation (1), we shall have

140 DIFFERENTIAL CALCULUS.

$$f(x+h, y+k) = u + \frac{du}{dy}k + \frac{d^2u}{dy^2}\frac{k^2}{1.2} + \frac{d^3u}{dy^3}\frac{k^3}{1.2.3} +, \text{etc.,}$$
$$+ \frac{du}{dx}h + \frac{d^2u}{dxdy}hk + \frac{d^3u}{dxdy^2}\frac{hk^2}{1.2} +, \text{etc.,}$$
$$+ \frac{d^2u}{dx^2}\frac{h^2}{1.2} + \frac{d^3u}{dx^2dy}\frac{h^2k}{1.2} +, \text{etc.,}$$

(2)
$$+ \frac{d^3u}{dx^3}\frac{h^3}{1.2.3} +, \text{etc.}$$

The general term being
$$\frac{d^{m+n}u}{dx^m dy^n} \cdot \frac{h^m . k^n}{(1.2\ldots m)(1.2\ldots n)}.$$

If in (1) we had made y to vary first, by giving to it an increment, k, we would have had, by Taylor's Theorem,

$$f(x, y+k) = u + \frac{du}{dy}k + \frac{d^2u}{dy^2}\frac{k^2}{1.2} + \frac{d^3u}{dy^3}\frac{k^3}{1.2.3} +, \text{etc.,} \quad (3)$$

where, as before, u, $\frac{du}{dy}$, $\frac{d^2u}{dy^2}$, etc., are functions of x and y.

If we give to x an increment, h,

u will become $u + \frac{du}{dx}h + \frac{d^2u}{dx^2}\frac{h^2}{1.2} + \frac{d^3u}{dx^3}\frac{h^3}{1.2.3} +, \text{etc.,}$

$\frac{du}{dy}$ " $\frac{du}{dy} + \frac{d^2u}{dydx}h + \frac{d^3u}{dydx^2}\frac{h^2}{1.2} +, \text{etc.,}$

$\frac{d^2u}{dy^2}$ " $\frac{d^2u}{dy^2} + \frac{d^3u}{dy^2dx}h + \frac{d^4u}{dy^2dx^2}\frac{h^2}{1.2} +, \text{etc.,}$

$\frac{d^3u}{dy^3}$ " $\frac{d^3u}{dy^3} + \frac{d^4u}{dy^3dx}h +, \text{etc.,}$

etc., etc.

Substituting these developments of u, $\frac{du}{dy}$, etc., in (3), we shall have the same development of the function,

$$f(x+h, y+k) = u + \frac{du}{dx}h + \frac{d^2u}{dx^2}\frac{h^2}{1.2} + \frac{d^3u}{dx^3}\frac{h^3}{1.2.3} +, \text{etc.,}$$
$$\frac{du}{dy}k + \frac{d^2u}{dydx}\frac{kh}{1.1} + \frac{d^3u}{dydx^2}\frac{kh^2}{1.2} +, \text{etc.,}$$
$$+ \frac{d^2u}{dy^2}\frac{k^2}{1.2} + \frac{d^3u}{dy^2dx}\frac{k^2h}{1.2} +, \text{etc.,}$$

(4)
$$+ \frac{d^3u}{dy^3}\frac{k^3}{1.2.3} +, \text{etc.}$$

The general term being
$$\frac{d^{n+m}u}{dx^n dy^m} \cdot \frac{h^n k^m}{(1.2\ldots n)(1.2\ldots m)},$$
equations (2) and (4) being identical, we have, by equating like powers of h and k,
$$\frac{d^2u}{dxdy} = \frac{d^2u}{dydx},$$
$$\frac{d^3u}{dxdy^2} = \frac{d^3u}{dy^2dx};$$
and generally, $\dfrac{d^{m+n}u}{dx^m dy^n} = \dfrac{d^{n+m}u}{dy^n dx^m}.$

We therefore conclude *that if a function of two independent variables is to be differentiated m times with respect to one of them, and n times with respect to the other, the result will be the same in whatever order the differentiations may be performed.*

If in (2) or (4) we transpose u, which is equal to $f(x, y)$, into the first member, we shall have

$$f(x+h, y+k) - f(x, y) = \frac{du}{dx}h + \frac{d^2u}{dx^2}\frac{h^2}{1.2} +, \text{ etc.,}$$
$$\frac{du}{dy}k + \frac{d^2u}{dydx}k \cdot h +, \text{ etc.,}$$
$$+ \frac{d^2u}{dy^2}\frac{k^2}{1.2} + \text{ etc. } \quad (5)$$

Then, passing to the limit,
$$du = \frac{du}{dx}dx + \frac{du}{dy}dy, \quad (6)$$

where the first member, du, denotes the *total* differential of the function; $\dfrac{du}{dx}dx$ denotes the *partial* differential of the function with respect to x; and $\dfrac{du}{dy}dy$ denotes the *partial* differential of the function with respect to y.

65. If we have a function of three variables, as
$$u' = f(x, y, z),$$
and suppose one of them, as z, to remain constant, giving to x an increment, h, and to y an increment, k, the development of the function
$$f(x+h, y+k, z)$$

will be of the same form as the development of
$$f(x+h, y+k);$$
but in this development we shall have u and all the differential coefficients, functions of x, y, and z. If we then take these several terms and give to z an increment, l, we shall have four terms of the development of the form
$$u, \frac{du}{dx}h, \frac{du}{dy}k, \frac{du}{dz}l.$$

Thus, if $\quad u'=f(x+h, y+k, z+l)$,

then $u'=u+\dfrac{du}{dz}l+\dfrac{d^2u}{dz^2}\dfrac{l^2}{1.2}+\dfrac{d^3u}{dz^3}\dfrac{l^3}{1.2.3}+$, etc.,

$\quad+\dfrac{du}{dy}k+\dfrac{d^2u}{dydz}l.k+\dfrac{d^3u}{dydz^2}\dfrac{l^2k}{1.2}+$, etc.,

$\quad+\dfrac{du}{dx}h+\dfrac{d^2u}{dxdz}l.h+\dfrac{d^3u}{dxdz^2}\dfrac{l^2h}{1.2}+$, etc.,

$\quad+\dfrac{d^2u}{dx^2}\dfrac{h^2}{1.2}+\dfrac{d^3u}{dx^2dz}\dfrac{h^2l}{1.2}+$, etc.,

$\quad+\dfrac{d^2u}{dxdy}hk+\dfrac{d^3u}{dxdydz}hkl+$, etc.,

$\quad+\dfrac{d^2u}{dy^2}\dfrac{k^2}{1.2}+\dfrac{d^3u}{dy^2dz}\dfrac{k^2l}{1.2}+$, etc.

Again, if we have a function of four variables, as
$$u=f(x, y, z, w),$$
there would be five terms of the development of the form $u, \dfrac{du}{dx}h, \dfrac{du}{dy}k, \dfrac{du}{dz}l, \dfrac{du}{dw}m$.

Every new variable that is introduced into the function will bring into the development a term containing the first power of the increment of that variable.

If, after we have obtained the development of a function of three variables, we transpose u, which is the original function, into the first member, and then pass to the limit, we shall have
$$d(f(x, y, z))=\frac{du}{dx}dx+\frac{du}{dy}dy+\frac{du}{dz}dz.$$
And similarly, for a function of four variables,
$$d(f(x, y, z, w))=\frac{du}{dx}dx+\frac{du}{dy}dy+\frac{du}{dz}dz+\frac{du}{dw}dw.$$

66. Hence we may conclude *that the total differential*

of a function of any number of variables is equal to the sum of all the partial differentials.

By the above *Art.* we can differentiate any algebraic function.

Ex. 130. Let $u = ax^3 + by^2 - cz$;

then $\dfrac{du}{dx} dx = 3ax^2 dx =$ first partial differentiation.

$\dfrac{du}{dy} dy = 2by\, dy =$ second "

$\dfrac{du}{dz} dz = -cdz =$ third "

$\therefore du = 3ax^2 dx + 2by\, dy - cdz.$

Ex. 131. Let $u = bx^2 + cy^3 - z$.
Ex. 132. Let $u = xy$.
Ex. 133. Let $u = xyz$.
Ex. 134. Let $u = x^2 y^3$.
Ex. 135. Let $u = x^4 y^4$.
Ex. 136. Let $u = x^m y^n$.

Ex. 137. Let $u = \dfrac{x}{y}$.

Ex. 138. Let $u = \dfrac{ay}{\sqrt{x^2 + y^2}}$.

Ex. 139. Let $u = vwxy$.
Ex. 140. Let $u = vwxyz$.

67. We shall now proceed to the successive differentials of a function of two variables, and the successive differential coefficients.

Resuming equation (6),

$$du = \dfrac{du}{dx} dx + \dfrac{du}{dy} dy.$$

Because $\dfrac{du}{dx}$ and $\dfrac{du}{dy} dy$ are both functions of x and y, each of these must be differentiated with respect to both variables, dx and dy being considered constant.

Taking the first differential, viz., $\dfrac{du}{dx} dx$, and differentiating with respect to x, we have

$d\left(\dfrac{du}{dx} dx\right)$ with respect to x becomes $\dfrac{d^2 u}{dx^2} dx^2$,

$d\left(\dfrac{du}{dx}dx\right)$ with respect to y becomes $\dfrac{d^2u}{dxdy}dxdy$.

Again, taking the second term, viz., $\dfrac{du}{dy}dy$,

$d\left(\dfrac{du}{dy}dy\right)$ with respect to x becomes $\dfrac{d^2u}{dydx}dydx$,

$d\left(\dfrac{du}{dy}dy\right)$ with respect to y becomes $\dfrac{d^2u}{dy^2}dy^2$.

But (Art. 64) $\dfrac{d^2u}{dxdy} = \dfrac{d^2u}{dydx}$;

$$\therefore d^2u = \dfrac{d^2u}{dx^2}dx^2 + 2\dfrac{d^2u}{dxdy}dxdy + \dfrac{d^2u}{dy^2}dy^2.$$

In order to obtain the third differential of the function, we must differentiate each one of these three terms twice; once with respect to x, and once with respect to y. Thus,

$d\left(\dfrac{d^2u}{dx^2}dx^2\right)$ with respect to x becomes $\dfrac{d^3u}{dx^3}dx^3$,

$d\left(\dfrac{d^2u}{dx^2}dx^2\right)$ with respect to y becomes $\dfrac{d^3u}{dx^2dy}dx^2dy$,

$d\left(2\dfrac{d^2u}{dxdy}dxdy\right)$ with respect to x becomes $2\dfrac{d^3u}{dx^2dy}dx^2dy$,

$d\left(2\dfrac{d^2u}{dxdy}dxdy\right)$ with respect to y becomes $2\dfrac{d^3u}{dxdy^2}dxdy^2$,

$d\left(\dfrac{d^2u}{dy^2}dy^2\right)$ with respect to x becomes $\dfrac{d^3u}{dy^2dx}dy^2dx$,

$d\left(\dfrac{d^2u}{dy^2}dy^2\right)$ with respect to y becomes $\dfrac{d^3u}{dy^3}dy^3$.

And (Art. 64) $\dfrac{d^3u}{dxdy^2} = \dfrac{d^3u}{dy^2dx}$;

$$\therefore d^3u = \dfrac{d^3u}{dx^3}dx^3 + 3\dfrac{d^3u}{dx^2dy}dx^2dy + 3\dfrac{d^3u}{dxdy^2}dxdy^2 + \dfrac{d^3u}{dy^3}dy^3.$$

By observing the analogy between the partial differentials and the terms of the development, we can easily find the subsequent differentials.

From the above, we see that *a function of two variables has two partial differential coefficients of the first order, three of the second, four of the third,* etc.

TRANSCENDENTAL FUNCTIONS.

68. *To find the differential of the logarithm of a quantity.*
Let $u = \log. x$,
and give to x an increment, h; then
$$u' = \log. (x+h),$$
and $\quad u' - u = \log. (x+h) - \log. x,$
$$= \log. \frac{x+h}{x} = \log. \left(1 + \frac{h}{x}\right).$$
But, by (Alg., Art. 164),
$$\log. \left(1 + \frac{h}{x}\right) = M\left[\frac{h}{x} - \tfrac{1}{2}\cdot\frac{h^2}{x^2} + \tfrac{1}{3}\cdot\frac{h^3}{x^3} - \tfrac{1}{4}\frac{h^4}{x^4} +, \text{etc.}\right],$$
where M denotes the modulus of the system;
$$\therefore \frac{u'-u}{h} = M\left[\frac{1}{x} - \frac{h}{2x^2} + \frac{h^2}{3x^3} - \frac{h^3}{4x^4} +, \text{etc.}\right].$$
The first member expresses the ratio of the increment of the function to that of the variable; and when we pass to the limit by making $h=0$, we have
$$\frac{du}{dx} = \frac{M}{x};$$
$$\therefore du = \frac{M dx}{x}, \text{ or } d.\log. x = \frac{M}{x}dx. \quad\text{(A)}$$
Hence *the differential of the logarithm of a quantity is equal to the modulus of the system into the differential of the quantity divided by the quantity itself.*
The modulus of the Naperian system is unity.

69. Therefore *the differential of a Naperian logarithm is equal to the differential of the quantity divided by the quantity itself.*
The Naperian logarithm is generally used in the Calculus, and when we wish to pass to the common system we have only to multiply the Naperian logarithm by the modulus of the common system.
Putting (Eq. A) into logarithms, we have
$$\log. d \log. x = \log. M + \log. dx - \log. x. \quad\text{(B)}$$
The modulus of the common system of logarithms is 0.434294, and the logarithm of this number is 9.637784.
Hence, to find the difference between the logarithms of two consecutive numbers, we employ (Eq. B), where dx is equal to *unity*. $\quad\therefore \log. dx = 0.$

G

Ex. 141. Find the differential of the common logarithm of 2563.

$$\begin{array}{r}\text{Log. modulus} \ldots \ldots 9.637784 \\ \text{Log. 2563} \ldots \ldots \ldots 3.408749 \\ \hline 6.229035\end{array}$$

The number corresponding to this logarithm is .000169, which is the difference between the logarithm of 2563 and the logarithm of 2564.

Examples in the Differentiation of Logarithmic Functions.

Ex. 142. Let $u = l\dfrac{1+x}{1-x}$.

Since the logarithm of a fraction is equal to the difference between the logarithm of the numerator and the logarithm of the denominator, we may put

$$u = l(1+x) - l(1-x).$$

Now, differentiating by the rule, we have

$$du = \frac{dx}{1+x} + \frac{dx}{1-x};$$

$$\therefore \frac{du}{dx} = \frac{2}{1-x^2}.$$

Ex. 143. Let $u = \log.^2 x = \log.$ of log. x.
If we put $\log. x = y$,
we shall have $u = \log. y$;
then $du = \dfrac{dy}{y}$.

But, since $y = \log. x$, $dy = \dfrac{dx}{x}$;

$$\therefore du = \frac{dx}{x \log. x}, \text{ or } \frac{du}{dx} = \frac{1}{x \log. x}.$$

Ex. 144. Let $u = \log.^3 x$.
Put $\log. x = z$, and $\log. z = y$;
then $\log.^2 x = \log. z = y$,
and $\log.^3 x = \log.^2 z = \log. y$,

$$\therefore u = \log. y, \; du = \frac{dy}{y}.$$

But $dy = \dfrac{dz}{z}$, and $dz = \dfrac{dx}{x}$.

TRANSCENDENTAL FUNCTIONS. 147

Substituting these, we have
$$du = \frac{dx}{x \cdot z \cdot y} = \frac{dx}{x \log. x \log.^2 x},$$
or
$$\frac{du}{dx} = \frac{1}{x \log. x \log.^2 x}.$$

Ex. 145. If $u = \log.^4 x$,
then
$$\frac{du}{dx} = \frac{1}{x \log. x \log.^2 x \log.^3 x}.$$
From which it is evident that if $u = \log.^n x$,
$$\frac{du}{dx} = \frac{1}{x \log. x \log.^2 x \ldots (\log.)^{n-1} x}.$$

Ex. 146. $u = l.\left(\dfrac{x}{\sqrt{a^2+x^2}}\right).$ $\quad \therefore \dfrac{du}{dx} = \dfrac{a^2}{x(a^2+x^2)}.$

Ex. 147. $u = l.(x + \sqrt{1+x^2}).$ $\quad \therefore \dfrac{du}{dx} = \dfrac{1}{\sqrt{1-x^2}}.$

Ex. 148. $u = l.\dfrac{a^{\frac{1}{2}} + x^{\frac{1}{2}}}{a^{\frac{1}{2}} - x^{\frac{1}{2}}}.$ $\quad \therefore \dfrac{du}{dx} = \dfrac{a^{\frac{1}{2}}}{x^{\frac{1}{2}}(a-x)}.$

Ex. 149. $u = l.\dfrac{x}{x + \sqrt{1+x^2}}.$ $\quad \therefore \dfrac{du}{dx} = \dfrac{1}{x} - \dfrac{1}{\sqrt{1+x^2}}.$

Ex. 150. $u = l.\dfrac{x}{a - \sqrt{a^2-x^2}}.$ $\quad \therefore \dfrac{du}{dx} = -\dfrac{a}{x\sqrt{a^2-x^2}}.$

Ex. 151. $u = l.\left(\dfrac{r + \sqrt{r^2-y^2}}{r - \sqrt{r^2-y^2}}\right).$ $\quad \therefore \dfrac{du}{dy} = -\dfrac{2r}{y\sqrt{r^2-y^2}}.$

Ex. 152. $u = l.\left(\dfrac{\sqrt{r+x} + \sqrt{r-x}}{\sqrt{r+x} - \sqrt{r-x}}\right).$
$$\therefore \frac{du}{dx} = -\frac{r}{x\sqrt{r^2-x^2}}.$$

70. *To find the differential of an exponential function.*
Let
$$u = a^x,$$
where a is a constant quantity.
Taking the logarithms of both members in the Naperian system, we shall have
$$\log. u = x \log. a;$$
and, differentiating by (Art. 69),
$$\frac{du}{u} = dx . \log. a,$$

or $$du = u \cdot \log. \, a \, dx.$$
Substituting for u its value a^x, we obtain
$$d \cdot a^x = a^x \log. \, a \, dx.$$
That is, *the differential of a constant raised to a variable exponent is equal to the quantity itself into the Naperian logarithm of the constant into the differential of its exponent.*

Taking the successive differential coefficients, we have
$$\frac{du}{dx} = a^x \log. \, a,$$
$$\frac{d^2u}{dx^2} = a^x (\log. \, a)^2,$$
$$\frac{d^3u}{dx^3} = a^x (\log. \, a)^3,$$
$$\frac{d^4u}{dx^4} = a^x (\log. \, a)^4.$$

Applying Maclaurin's Theorem, we shall have, by making $x = 0$,
$$U = 1, \; U' = \log. \, a, \; U'' = (\log. \, a)^2, \; U''' = (\log. \, a)^3,$$
$$U^{iv} = (\log. \, a)^4 ;$$
$$\therefore a^x = 1 + x . \log. \, a + \tfrac{1}{2}(x . \log. \, a)^2 + \frac{1}{2 \cdot 3}(x \log. \, a)^3 +, \text{etc.} \quad (A)$$

If we make $x = 1$, we shall have
$$a = 1 + \log. \, a + \tfrac{1}{2}(\log. \, a)^2 + \frac{1}{2 \cdot 3}(\log. \, a)^3 +, \text{etc.} \quad (B)$$

In (Eq. A), by changing a into the Naperian base e, we have
$$e^x = 1 + x + \tfrac{1}{2}x^2 + \frac{1}{2 \cdot 3}x^3 +, \text{etc.,}$$
because the $\log. \, e = 1$.

71. Now if $x = 1$, we obtain
$$e = 1 + 1 + \tfrac{1}{2} + \frac{1}{1 \cdot 2 \cdot 3} + \frac{1}{1 \cdot 2 \cdot 3 \cdot 4} +, \text{etc.} = 2.71828 +,$$
which is the base of the Naperian system of logarithms.

72. As the logarithms of the same number, taken in two different systems, are to each other as the moduli of those systems, therefore,

> As log. e in the Naperian system
> : log. e in the common system
> :: modulus of the Naperian system
> : modulus of the common system.

TRANSCENDENTAL FUNCTIONS.   149

But the log. of e in the Nap. system is *unity*, and in the common system it is 0.434294; hence
$$1 : 0.434294 :: 1 : 0.434294 = \text{mod. com. sys.}$$
Again: *to find the Naperian logarithm of ten*, we say,
As the modulus of the common system
: the modulus of the Naperian system
:: the log. of 10 in the common system
: the log. of 10 in the Naperian system.
That is, as $0.434294 : 1 :: 1 : 2.302585 = \log. 10$ in the Naperian system.

73. Let $\qquad u = x^y.$
Taking the logarithm of both members, we have
$$\log. u = y \log. x;$$
and differentiating,
$$\frac{du}{u} = \log. x \, dy + y \frac{dx}{x},$$
or $\qquad du = u \log. x \, dy + uy \frac{dx}{x};$
$$\therefore \log. x^y = x^y \log. x \, dy + yx^{y-1} . dx.$$
Hence, *to differentiate a variable root raised to a variable exponent, differentiate first under the supposition that the root is constant, then that the exponent is constant, and take the sum of these differentials.*

Ex. 153. Let $\qquad u = x^{y^z},\qquad$ (1)
where x, y, and z are variables.
Put $\qquad w = y^z;\qquad$ (2)
then $\qquad dw = zy^{z-1}dy + y^z \log. y \, dz.\qquad$ (3)
But $\qquad u = x^w;$
$$\therefore du = wx^{w-1}dx + x^w \log. x \, dw.\qquad (4)$$
Substitute in (4) the values of w and dw, and reduce, we obtain
$$du = x^{y^z} y^z \left(l y l x \, dz + \frac{zl . x \, dy}{y} + \frac{dx}{x} \right).$$
Ex. 154. Let $u = x^{x^x}.$
$$\therefore \frac{du}{dx} = x^{x^x} . x^x \left(l . x (l . x + 1) + \frac{1}{x} \right).$$
Ex. 155. Let $u = x^x.$
$$\therefore \frac{du}{dx} = x^x (\log. x + 1).$$

Ex. 156. Let $u = a^{\log. x}$.
$$\therefore \frac{du}{dx} = \frac{a^{\log. x} . \log. a}{x}.$$

Ex. 157. Let $u = x^{\frac{1}{x}}$.
$$\therefore \frac{du}{dx} = \frac{1}{x^2} . x^{\frac{1}{x}} \log. \left(\frac{e}{x}\right).$$

Ex. 158. Let $u = (lx)^n$.
$$\therefore \frac{du}{dx} = \frac{n(lx)^{n-1}}{x}.$$

CIRCULAR FUNCTIONS.

74. *The differential coefficient of the sine of an arc, radius being* (1), *is equal to the cosine of the arc.*

Let $u = \sin. x$,
and give to x an increment, h;
then $u' = \sin. (x+h)$,
and $u' - u = \sin. (x+h) - \sin. x$.
But, by (Trig., Eq. 11),
$$\sin. (x+h) - \sin. x = 2 \sin. \tfrac{1}{2}h \cos. (x+\tfrac{1}{2}h);$$
$$\therefore u' - u = 2 \sin. \tfrac{1}{2}h . \cos. (x+\tfrac{1}{2}h),$$
and
$$\frac{u'-u}{h} = \frac{2 . \sin. \tfrac{1}{2}h}{h} . \cos. (x+\tfrac{1}{2}h).$$

Divide the numerator and denominator of the fraction in the second member by 2,
$$\frac{u'-u}{h} = \frac{\sin. \tfrac{1}{2}h}{\tfrac{1}{2}h} \cos. (x+\tfrac{1}{2}h).$$

Passing to the limit,
$$\frac{du}{dx} = \frac{d . \sin. x}{dx} = \cos. x,$$
or $d . \sin. x = \cos. dx$.

75. *The differential coefficient of the cosine of an arc, radius being* (1), *is negative, and equal to the sine of the arc.*

Let $u = \cos. x$;
then $u' = \cos. (x+h)$,
$u' - u = \cos. (x+h) - \cos. x$.
But, by (Trig., Eq. 14),
$$\cos. (x+h) - \cos. x = 2 \sin. (x+\tfrac{1}{2}h) \sin. (-\tfrac{1}{2}h) =$$
$$-2 \sin. \tfrac{1}{2}h \sin. (x+\tfrac{1}{2}h).$$

CIRCULAR FUNCTIONS.   151

$$\therefore u'-u = -2\sin.\tfrac{1}{2}h\sin.(x+\tfrac{1}{2}h),$$

$$\frac{u'-u}{h} = \frac{-2\sin.\tfrac{1}{2}h}{h}\sin.(x+\tfrac{1}{2}h),$$

$$= \frac{-\sin.\tfrac{1}{2}h}{\tfrac{1}{2}h}\sin.(x+\tfrac{1}{2}h);$$

$$\therefore \frac{du}{dx} = -\sin. x, \text{ or } d.\cos. x = -\sin. x\, dx.$$

Because versin. $x = 1 - \cos. x$, the diff. of versin. $x = -$diff. of cos. $x = \sin. x\, d.x$.

76. *The differential coefficient of the tangent of an arc, the radius being* (1), *is equal to the square of the secant, or* $= (1+\tan.^2)$.

Let
$$u = \tan. x,$$
$$u' = \tan.(x+h),$$
$$u'-u = \tan.(x+h) - \tan. x,$$
$$= \frac{\sin.(x+h)}{\cos.(x+h)} - \frac{\sin. x}{\cos. x},$$
$$= \frac{\sin. h}{\cos.(x+h)\cos. x};$$
$$\frac{u'-u}{h} = \frac{\dfrac{\sin. h}{h}}{\cos.(x+h)\cos. x}.$$
$$\frac{du}{dx} = \frac{1}{\cos.^2 x} = \sec.^2 x = 1 + \tan.^2 x,$$
$$du = \sec.^2 x\, dx = (1+\tan.^2 x)dx.$$

77. *The differential coefficient of the cotangent of an arc, radius being unity, is negative, and equal to the square of the cosecant, or* $= -(1+\cot.^2)$.

Let $u = \cot. x$;
then $u' = \cot.(x+h),$
and $u'-u = \cot.(x+h) - \cot. x,$
$$= \frac{\cos.(x+h)}{\sin.(x+h)} - \frac{\cos. x}{\sin. x},$$
$$= -\frac{\sin. h}{\sin.(x+h)\sin. x},$$
$$\frac{u'-u}{h} = -\frac{\dfrac{\sin. h}{h}}{\sin.(x+h)\sin. x},$$

$$\frac{du}{dx} = -\frac{1}{\sin.^2 x} = -\text{cosec.}^2 x,$$
$$= -(1 + \cot.^2 x);$$
$$\therefore du = -(1 + \cot.^2 x) dx.$$

78. *The differential coefficient of the secant of an arc, radius being* (1), *is equal to the rectangle of the secant and tangent.*

Let $u = \sec. x$;
then $u' = \sec. (x+h),$
and $u' - u = \sec. (x+h) - \sec. x,$
$$= \frac{1}{\cos. (x+h)} - \frac{1}{\cos. x},$$
$$= \frac{\cos. x - \cos. (x+h)}{\cos. (x+h) \cos. x}.$$

But, by Plane Trig., Eq. 14,
$$\cos. x - \cos. (x+h) = 2 \sin. (x+\tfrac{1}{2}h) \sin. \tfrac{1}{2}h;$$
$$\therefore u' - u = \frac{2 \sin. (x+\tfrac{1}{2}h) \sin. \tfrac{1}{2}h}{\cos. (x+h) \cos. x},$$
$$\frac{u'-u}{h} = \frac{\sin. (x+\tfrac{1}{2}h)}{\cos. (x+h) \cos. x} \cdot \frac{\sin. \tfrac{1}{2}h}{\tfrac{1}{2}h};$$
$$\therefore \frac{du}{dx} = \frac{\sin. x}{\cos.^2 x} = \frac{\sin. x}{\cos. x} \cdot \frac{1}{\cos. x} = \sec. x \tan. x,$$
or $du = \sec. x \tan. x \, dx.$

79. *The differential coefficient of the cosecant of an arc, radius being* (1), *is negative, and equal to the rectangle formed by the cosecant and cotangent of the arc.*

Let $u = \text{cosec.} x$;
then $u' = \text{cosec.} (x+h),$
and $u' - u = \text{cosec.} (x+h) - \text{cosec.} x,$
$$= \frac{1}{\sin. (x+h)} - \frac{1}{\sin. x},$$
$$= \frac{\sin. x - \sin. (x+h)}{\sin. (x+h) \sin. x}.$$

But, by Plane Trig., Eq. 12,
$$\sin. x - \sin. (x+h) = -2 \sin. \tfrac{1}{2}h \cos. (x+\tfrac{1}{2}h);$$
$$\therefore u' - u = -\frac{2 \sin. \tfrac{1}{2}h \cos. (x+\tfrac{1}{2}h)}{\sin. (x+h) \sin. x},$$
$$\frac{u'-u}{h} = -\frac{\sin. \tfrac{1}{2}h}{\tfrac{1}{2}h} \cdot \frac{\cos. (x+\tfrac{1}{2}h)}{\sin. (x+h) \sin. x};$$

CIRCULAR FUNCTIONS. 153

$$\therefore \frac{du}{dx} = -\frac{\cos. x}{\sin.^2 x} = -\frac{\cos. x}{\sin. x} \cdot \frac{1}{\sin. x} = -\cot. x \cosec. x,$$

$d . \cosec. x = -\cosec. x \cot. x \, dx.$

In order to find the differential of the *natural sine*, etc., of an arc, the differences being taken for single minutes, we must use the equations in Arts. 74–79, after having made them homogeneous. Thus:

Ex. 159. Find the differential of the natural sine of 18° 30'.

Art. 74, $d \sin. x = \dfrac{\cos. x \, dx}{R},$

or log. $d \sin. x = $ log. cos. $x + $ log. $dx - 10$.

The length of the sine of one minute is (Trig., Art. 10), 0.00029088, which is equal to dx.

Hence $x = 18° 30'$, log. cos. $= 9.976957$
   $dx = .00029088$, its log. $= 6.463726$
   $\overline{6.440683}$

The number corresponding to this logarithm is
   .000275,

which is the difference between the natural sine of 18° 30' and that of 18° 31'.

Ex. 160. Find the differential of the natural cosine of 18° 30'.

Art. 75, $-d \cos. x = \dfrac{\sin. x \, dx}{R}.$

Hence $x = 18° 30'$, log. sin. $= 9.501476$
   dx log. $= 6.463726$
   $-$log. $d \cos. x$ $\overline{5.965202}$

The number corresponding to this logarithm is
   $-.000092,$

which is the difference between the natural cosine of 18° 30' and that of 18° 31', and being negative, it shows that it is to be subtracted from the cosine of 18° 30' to obtain the cosine of 18° 31'.

80. Having gone through the differentials of the *natural* sines, tangents, etc., we shall proceed with the differentials of the *logarithmic* sines, etc.

Let $u = $ log. sin. $x;$ then $du = M \cdot \dfrac{\cos. x}{\sin. x} \cdot dx,$

$\therefore d . $ log. sin. $x = M . \cot. x . dx.$

G 2

Hence *the differential of the logarithmic sine of an arc* (rad.$=1$) *is equal to the modulus of the system into the cotangent of the arc into the differential of the arc.*

81. *The differential of the logarithmic cosine of an arc* (rad.$=1$) *is negative, and equal to the modulus of the system into the tangent of the arc into the differential of the arc.*

Let $\quad u = \log. \cos. x$;

then $\quad du = -M \cdot \dfrac{\sin. x}{\cos. x} \cdot dx.$

$\therefore d.\log.\cos. x = -M \cdot \tan. x \cdot dx.$

82. Let $\quad u = \log. \tan. x$;

then $\quad du = \dfrac{M(1+\tan.^2 x)dx}{\tan. x}$;

$\therefore d.\log.\tan. x = M \cdot \dfrac{\sec.^2 x}{\tan. x} \cdot dx = \dfrac{Mdx}{\sin. x \cos. x}.$

83. Let $\quad u = \log. \cot. x$;

then $\quad du = \dfrac{-M(1+\cot.^2 x)dx}{\cot. x},$

$\qquad = \dfrac{-M \cosec.^2 x \, dx}{\cot. x} = \dfrac{-M \cdot dx}{\sin. x \cos. x}.$

Hence *the differentials of the log. tan. and log. cotan. of any arc are equal with contrary signs.*

To find the differential of the *logarithmic sine*, etc., of an arc, the differences being taken in seconds, we use the equations (Art. 80–83). The length of one second is .00000485 (Trig., Art. 10).

Ex. 161. Find the differential of log. sine $18° \, 30'$.

$\quad$ log. modulus $\ldots = 9.637784$
$\quad x = 18° \, 30'$ log. cot. $\,. \,. = 10.475480$
$\quad dx = .00000485$, its log. $. \,. = 4.685742$
$\qquad\qquad$ log. d log. sin. $x = \overline{4.799006}$

The natural number corresponding to this is
$\qquad\qquad$ .000006,
which is the difference between the log. sines of $18° \, 30'$ and $18° \, 30' \, 1''$.

84. In treating the circular functions, it is sometimes more convenient to regard the arc as the function, and the sine, cosine, etc., as the variable.

CIRCULAR FUNCTIONS.   155

If we designate the variable by u, and the arc by x, we shall have
$$x = \sin.^{-1} u,$$
which is read x equals an arc whose sine is u. See Trig., Chap. IV.
Then $\quad\sin. x = u,$
and, by differentiation,
$$\cos. x \, dx = du;$$
$$\therefore dx = \frac{du}{\cos. x}.$$
But $\quad \cos. x = \sqrt{1 - \sin.^2 x}$, and $\sin.^2 x = u^2$;
$$\therefore dx = \frac{du}{\sqrt{1-u^2}}.$$

85. Let $\quad x = \cos.^{-1} u;$
then $\quad \cos. x = u,$
and, by differentiating,
$$-\sin. x \, dx = du;$$
$$\therefore dx = \frac{-du}{\sin. x}.$$
But $\quad \sin. x = \sqrt{1 - \cos.^2 x}$, and $\cos.^2 x = u^2$;
$$\therefore dx = \frac{-du}{\sqrt{1-u^2}}.$$

86. Let $\quad x = \tan.^{-1} u;$
then $\quad \tan. x = u,$
and, by differentiating,
$$(1 + \tan.^2 x) dx = du;$$
$$\therefore dx = \frac{du}{1 + \tan.^2 x}, \text{ and } \tan.^2 x = u^2;$$
$$\therefore dx = \frac{du}{1 + u^2}.$$

87. If $\quad x = \cot.^{-1} u, \therefore dx = -\frac{du}{1 + u^2}.$

88. Let $\quad x = \text{versin.}^{-1} u;$
then $\quad \text{versin.} x = u,$
and, by differentiating,
$$\sin. x \, dx = du;$$
$$\therefore dx = \frac{du}{\sin. x}.$$

156 DIFFERENTIAL CALCULUS.

But $\text{versin. } x = 1 - \cos. x = 1 - \sqrt{1 - \sin.^2 x}$,
$\sqrt{1 - \sin.^2 x} = 1 - \text{versin. } x = 1 - u$,
$1 - \sin.^2 x = 1 - 2u + u^2$,
$\sin. x = \sqrt{2u - u^2}$;
$$\therefore dx = \frac{du}{\sqrt{2u - u^2}}.$$

89. Let $x = \sec.^{-1} u$;
then $\sec. x = u$,
and, by differentiating,
$\sec. x \tan. x \, . \, dx = du$;
$$\therefore dx = \frac{du}{\sec. x \, . \, \tan. x},$$
$$= \frac{du}{u\sqrt{u^2 - 1}}.$$

90. In like manner, if $x = \text{cosec.}^{-1} u$,
then $$dx = -\frac{du}{u\sqrt{u^2 - 1}}.$$

91. If s be an arc whose sine is x, we shall have, from Art. 84, $s = \sin.^{-1} x$;
then $$\frac{dx}{ds} = \cos. s.$$

Formulas (84), (85), (86), (87), (88), (89), (90), being of frequent use, should be thoroughly committed to memory.

Examples.

Ex. 162. Differentiate the function
$$u = \sin.^{-1} (2x\sqrt{1 - x^2}).$$
$$\therefore \frac{du}{dx} = \frac{2}{\sqrt{1 - x^2}}.$$

Ex. 163. If $u = \dfrac{2}{\sqrt{3}} \tan.^{-1} \left[\dfrac{1 + 2x}{\sqrt{3}} \right]$.
$$\therefore \frac{du}{dx} = \frac{1}{1 + x + x^2}.$$

Ex. 164. If $u = \cos.^{-1} (x\sqrt{1 - x^2})$.
$$\therefore \frac{du}{dx} = \frac{(-1 + 2x^2)}{\sqrt{1 - x^2 + x^4(1 - x^2)}}.$$

Ex. 165. If $u = \tan.^{-1} \dfrac{x}{y}$,

$$\frac{du}{dx} = \frac{y - x\dfrac{dy}{dx}}{y^2 + x^2}.$$

92. Methods of expansion are often adopted in special cases, of which we shall give some examples.

To expand $\sin.^{-1} x$ in powers of x.

Assume $\sin.^{-1} x = A_0 + A_1 x + A_2 x^2 +$, etc. (1)

Differentiating both members of this equation, and dividing through by dx, we have

$$\frac{1}{\sqrt{1-x^2}} = A_1 + 2A_2 x + 3A_3 x^2 +, \text{ etc.} \quad (2)$$

But, by the Binomial Theorem,

$$\frac{1}{\sqrt{1-x^2}} = 1 + \tfrac{1}{2}x^2 + \frac{1 \cdot 3}{2 \cdot 4}x^4 + \frac{1 \cdot 3 \cdot 5}{2 \cdot 4 \cdot 6}x^6 +, \text{ etc.} \quad (3)$$

Comparing the coefficients in (2) and (3), we have, by first making $x = 0$ in (1),

$$A_0 = 0, \ A_1 = 1, \ A_2 = 0, \ A_3 = \frac{1}{2 \cdot 3}, \text{ etc.;}$$

$$\therefore \sin.^{-1} x = x + \frac{x^3}{2 \cdot 3} + \frac{3x^5}{2 \cdot 4 \cdot 5} +, \text{ etc.}$$

2. To express an arc in terms of its tangent:

Assume $\tan.^{-1} x = A_0 + A_1 x + A_2 x^2 + A_3 x^3 +$, etc. (1)

Differentiating both members, and dividing by dx, we have

$$\frac{1}{1+x^2} = A_1 + 2A_2 x + 3A_3 x^2 +, \text{ etc.} \quad (2)$$

But

$$\frac{1}{1+x^2} = 1 - x^2 + x^4 - x^6 + x^8 +, \text{ etc.} \quad (3)$$

Equating coefficients of like powers of x in (2) and (3), we have

$$A_1 = 1, \ A_2 = 0, \ A_3 = -\tfrac{1}{3}, \text{ etc.}$$

And putting $x = 0$ in (1), we get $A_0 = 0$;

$$\therefore \tan.^{-1} x = x - \frac{x^3}{3} + \frac{x^5}{5} - \frac{x^7}{7} +, \text{ etc.,}$$

$$= x\left(1 - \frac{x^2}{3} + \frac{x^4}{5} - \frac{x^6}{7} + \frac{x^8}{9} -, \text{ etc.}\right).$$

From this we can readily calculate the length of an arc of the circle; for if we call the arc whose tangent is x

$30°$, its tangent x will be equal to $\dfrac{1}{\sqrt{3}} = .57735$, and these being substituted in the above, we shall have

$$\text{arc } 30° = \tan.^{-1}\dfrac{1}{\sqrt{3}}$$

$$= \dfrac{1}{\sqrt{3}}\left(1 - \dfrac{1}{3.3} + \dfrac{1}{5.3^2} - \dfrac{1}{7.3^3} + \dfrac{1}{9.3^4} -, \text{etc.}\right),$$

$= .5235987 =$ the length of an arc of $30°$.

If we multiply this by 6, we shall have the semicircumference to radius unity $= 3.14159$.

93. By means of Maclaurin's Theorem we are enabled to find the value of the principal functions of an arc in terms of the arc itself. Thus:

Let $\qquad u = f(x) = \sin. x.$ \hfill (1)

If we make $x = 0$, then $U = 0$.

Differentiating (1), we have

$\qquad \dfrac{du}{dx} = \cos. x$, whence, if $x = 0$, $U' = 1$,

and $\qquad \dfrac{d^2u}{dx^2} = -\sin. x$, whence, if $x = 0$, $U'' = 0$,

and $\qquad \dfrac{d^3u}{dx^3} = -\cos. x$, whence, if $x = 0$, $U''' = -1$,

and $\qquad \dfrac{d^4u}{dx^4} = \sin. x$, whence, if $x = 0$, $U^{iv} = 0$,

and $\qquad \dfrac{d^5u}{dx^5} = \cos. x$, whence, if $x = 0$, $U^v = 1$

$$\therefore \sin. x = x - \dfrac{x^3}{1.2.3} + \dfrac{x^5}{1.2.3.5} -, \text{etc.}$$

94. To develop cosine x in terms of x.

Let $\qquad u = f(x) = \cos. x.$ \hfill (1)

If we make $x = 0$, $\cos. x = 1$; $\therefore U = 1$.

Differentiating (1) successively, and making x in each differential coefficient equal to zero, we have

$\qquad \dfrac{du}{dx} = -\sin. x\,;\ \therefore U' = 0.$

$\qquad \dfrac{d^2u}{dx^2} = -\cos. x\,;\ \therefore U'' = -1.$

$\qquad \dfrac{d^3u}{dx^3} = \sin. x\,;\ \therefore U''' = 0.$

IMPLICIT FUNCTIONS. 159

$$\frac{d^4u}{dx^4}=\cos. x\,; \quad \therefore U^{iv}=1.$$

etc., etc.;

$$\therefore \cos. x = 1 - \frac{x^2}{1.2} + \frac{x^4}{1.2.3.4} -, \text{etc.}$$

The formulas for the sine and cosine of an arc are very convenient in calculating tables, especially when the arc is small.

IMPLICIT FUNCTIONS.

95. When the relation between a function and its variable is expressed by the equation

$$f(x, y) = 0,$$

y may be called an implicit function of x, or x an implicit function of y. For if the equation $f(x, y) = 0$ can be solved with respect to either of them, the result may be

$$y = f(x),$$
or $$x = f'(y).$$

But, as it is often difficult to solve the given equation, it becomes necessary to obtain a rule for determining the value of $\frac{dy}{dx}$ which does not require this labor.

Put $\quad u = f(x, y) = 0,\quad$ (1)

and let x become $x+h$, and y become $y+k$, so that

$$f(x+h, y+k) = 0. \quad (2)$$

Hence $\quad f(x+h, y+k) - f(x, y) = 0. \quad (3)$

If we subtract from, and add to the first member of (3) the quantity $f(x+h, y)$, we shall have

$$f(x+h, y+k) - f(x+h, y) + f(x+h, y) - f(x, y) = 0. \quad (4)$$

Let us divide by h, and at the same time multiply the first fraction by $\frac{k}{k}$, and we shall have

$$\frac{f(x+h, y+k) - f(x+h, y)}{k} \cdot \frac{k}{h} + \frac{f(x+h, y) - f(x, y)}{h} = 0.$$
(5)

(Eq. 5) being always true, remains true when h and k are diminished indefinitely.

The limit of $\dfrac{f(x+h, y+k) - f(x+h, y)}{k}$ would be the differential coefficient of $f(x+h, y)$ *formed on the suppo-*

sition that y alone varies if h remained constant; but as h diminishes without limit when k diminishes without limit, the limit of the expression is the differential coefficient of $f(x, y)$, or u with respect to y formed on the supposition that y alone varies.

It may be represented by $\left(\dfrac{du}{dy}\right)$.

The limit of $\dfrac{k}{h}$ is $\dfrac{dy}{dx}$.

The limit $\dfrac{f(x+h, y) - f(x, y)}{h}$, when $h=0$, is the differential coefficient of $f(x, y)$, or *u with respect to x formed on the supposition that x alone varies*, and the limit may be denoted by $\left(\dfrac{du}{dx}\right)$. Therefore

$$\left(\frac{du}{dy}\right)\frac{dy}{dx} + \left(\frac{du}{dx}\right) = 0, \qquad (6)$$

and

$$\frac{dy}{dx} = -\frac{\left(\dfrac{du}{dx}\right)}{\left(\dfrac{du}{dy}\right)}. \qquad (7)$$

96. Hence *the differential coefficient of y regarded as a function of x is equal to the ratio of the partial differential coefficients of u regarded as a function of x, and u regarded as a function of y, taken with an opposite sign.*

Examples.

Ex. 166. Let $f(x, y) = u = y^2 + x^2 - R^2 = 0$; (1)

then $\left(\dfrac{du}{dx}\right) = 2x$, and $\left(\dfrac{du}{dy}\right) = 2y$;

$$\therefore \frac{dy}{dx} = -\frac{\left(\dfrac{du}{dx}\right)}{\left(\dfrac{du}{dy}\right)} = -\frac{x}{y}.$$

Differentiating the second time, we obtain

$$\frac{d^2y}{dx^2} = -\frac{x^2+y^2}{y^3} = -\frac{R^2}{y^3}.$$

APPLICATION OF THE CALCULUS TO CURVES. 161

Ex. 167. Let $u = y^2 + 2xy + x^2 - a^2 = 0$;

then $\dfrac{dy}{dx} = -1.$

Ex. 168. Let $u = y^2 - 2mxy + x^2 - a^2 = 0$;

then $\dfrac{dy}{dx} = \dfrac{my - x}{y - mx}.$

Ex. 169. Let $u = x^3 + 3axy + y^3 = 0$;

then $\dfrac{dy}{dx} = -\dfrac{x^2 + ay}{ax + y^2},$

and $\dfrac{d^2y}{dx^2} = -\dfrac{2a^3xy}{(ax + y^2)^3}.$

CHAPTER III.

APPLICATION OF THE DIFFERENTIAL CALCULUS TO THE THEORY OF CURVES.

97. We shall now apply the Calculus to Geometry, and first show how an equation may be freed from its constants.

For instance, if we take the equation of a straight line,

$$y = ax + b, \qquad (1)$$

and differentiate it, we shall have

$$\dfrac{dy}{dx} = a. \qquad (2)$$

Differentiating again, we find

$$\dfrac{d^2y}{dx^2} = 0. \qquad (3)$$

Equation (3) is entirely independent of the constants a and b; hence it is equally applicable to every straight line which can be drawn in the plane of the co-ordinate axes. It is called *the differential equation of lines of the first order*.

98. If we take the equation of the circle

$$x^2 + y^2 = R^2, \qquad (1)$$

and differentiate it, we shall have

$$\dfrac{dy}{dx} = -\dfrac{x}{y}. \qquad (2)$$

Equation (2) is independent of R, and hence it belongs to every circle whose centre is at the origin of rectangular co-ordinates.

Again: if we take the equation of the parabola
$$y^2 = 2px,$$
and differentiate it, we obtain
$$\frac{dy}{dx} = \frac{y}{2x}.$$

This equation is independent of the value of the parameter, and therefore belongs to every parabola referred to the same co-ordinate axes.

99. If we take the general equation of lines of the second order, which is
$$y^2 = mx + nx^2, \tag{1}$$
and differentiate it, we obtain
$$2y\,dy = m\,dx + 2nx\,dx. \tag{2}$$
Differentiating again, regarding dx as constant, we have
$$dy^2 + y\,d^2y = n\,dx^2. \tag{3}$$
Eliminating m and n from these three equations, we obtain
$$y^2 dx^2 + x^2 dy^2 + x^2 y\, d^2y - 2xy\,dx\,dy = 0,$$
which is *the general differential equation of lines of the second order*.

100. Hence, to free an equation of its constants, it will be necessary to differentiate it as many times as there are constants to be eliminated. We shall then have the original equation, together with the differential equations obtained, being one more in number than the constants. From these equations the constants may be eliminated by any one of the methods used in Algebra.

The differential equation which is obtained after the constants are eliminated belongs to a *species* or *order of lines*, one of which is represented by the given equation.

101. The differential equation of a species expresses the law by which the variable co-ordinates change their values, and this equation should therefore be independent of the constants, which determine the *magnitude*, and not the *nature* of the curve.

102. To determine the tangent of the angle which a tangent line to a curve makes with the axis of abscissas.

Let P be any point of the curve CPP' through which

APPLICATION OF THE CALCULUS TO CURVES. 163

a tangent line is to be drawn. We are required to find the tangent of the angle which this tangent line makes with the axis of x.

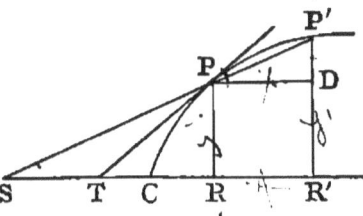

Let x and y denote the co-ordinates of the point P.

Give to x an increment, h, equal to RR'; draw R'P' perpendicular to CR, PD parallel to CR; also draw the secant SPP'.

Put
$$R'P' = y';$$
then
$$P'D = P'R' - DR = y' - y.$$
But in the triangle PDP' we have
$$\frac{P'D}{PD} = \text{tan. P'PD} = \text{tan. S}.$$
Substituting for P'D and PD their values, we obtain
$$\frac{y' - y}{h} = \text{tan. S}. \qquad (1)$$

The first member of (1) expresses the ratio of the increment of y to that of x. Now if h is diminished without limit, the point P' will approach P, the secant SPP' will approach the tangent TP; and when $h = 0$, the point P' will coincide with the point P, and the secant will coincide with the tangent. The first member of the equation becomes $\frac{dy}{dx}$, and the angle S becomes the angle T.

Therefore
$$\frac{dy}{dx} = \text{tan. T}.$$

Hence the differential coefficient of the ordinate is the tangent of the angle which a tangent line to the curve makes with the axis of x.

103. If it be required to determine the point of a curve at which a tangent line makes a *given* angle with the axis of x, *we must make the first differential coefficient of y, regarded as a function of x, equal to the tangent of the given angle.*

Examples.

Ex. 170. Find the co-ordinates of the points at which a tangent line to the circumference of a circle shall be

perpendicular to the axis of abscissas from the equation
$$y^2+x^2=R^2. \qquad (1)$$
By differentiating, we have
$$\frac{dy}{dx}=-\frac{x}{y}.$$
And as the tangent is to be perpendicular to the axis of x, the angle formed is a right angle, and its tangent is equal to *infinity*.
$$\therefore \frac{dy}{dx}=-\frac{x}{y}=\infty.$$
Hence $y=0$.
Substituting this value of y in (1), we have
$$x=\pm R.$$
The points at which a tangent line is perpendicular to the axis of x are those in which the circumference cuts that axis.

If it be required to find the points at which the tangent line is to be parallel to the axis of x, we must make
$$\frac{dy}{dx}=0; \quad \therefore x=0, \text{ and } y=\pm R.$$

Ex. 171. Find the points at which a tangent line to the ellipse is perpendicular to the axis of x from each of the equations
$$A^2y^2+B^2x^2=A^2B^2, \qquad (1)$$
and
$$y^2=\frac{B^2}{A^2}(2Ax-x^2), \qquad (2)$$
and also the points at which the tangent line is parallel to the axis of x.

Ex. 172. Find the point on a parabola at which a tangent line makes an angle of 45° with the axis.

Ex. 173. Find the abscissa of the point at which a tangent line to the curve is parallel to the axis of x, the equation of the curve being
$$y(x-1)(x-2)=x-3.$$
Ans. $x=3\pm\sqrt{2}.$

Ex. 174. Find the abscissa of the point at which a tangent is parallel to the axis of x, the equation of the curve being
$$y^3=(x-a)^2(x-c).$$
Ans. $x=\dfrac{2c+a}{3}.$

APPLICATION OF THE CALCULUS TO CURVES. 165

104. *To determine the length of the subtangent to any point of a curve referred to rectangular axes.*

In the right-angled triangle TRP, we have

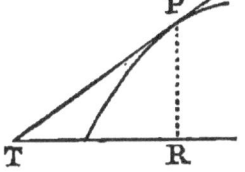

$$\frac{PR}{TR} = \tan T,$$

or $\quad \dfrac{y}{TR} = \dfrac{dy}{dx};$

$$\therefore TR = y \cdot \frac{dx}{dy} = \text{subtangent}.$$

105. Hence, to find the subtangent to any point of a curve, we have the following

RULE.

Differentiate the equation of the curve, and find the value of $y\dfrac{dx}{dy}$.

Examples.

Ex. 175. Find the subtangent of the curve whose equation is
$$y = \frac{x^{\frac{3}{2}}}{\sqrt{2a-x}}.$$

$$\text{Subtan.} = \frac{x(2a-x)}{3a-x}.$$

Ex. 176. Find the subtangent of the curve whose equation is $\quad y^3 - 3axy + x^3 = 0.$

$$\text{Subtan.} = \frac{2axy - x^3}{ay - x^2}.$$

Ex. 177. Find the subtangent of the curve whose equation is $\quad xy^2 = a^2(a-x).$

$$\text{Subtan.} = -\frac{2(ax - x^2)}{a}.$$

106. *To determine the length of the subnormal to any point of a curve.*

In the adjoining figure, RN is the subnormal, and it is required to find its length. In the right-angled triangle RPN, we have

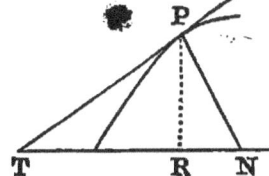

$$\frac{RN}{RP} = \tan NPR.$$

But the angle NPR is equal to NTP;

$$\therefore \frac{RN}{RP} = \frac{dy}{dx}; \text{ and as } RP = y,$$

$$\therefore RN = y\frac{dy}{dx} = \text{subnormal}.$$

107. Hence, to find the subnormal to any point of a curve referred to rectangular axes, we have the following

RULE.

Differentiate the equation of the curve, and find the value of $y\frac{dy}{dx}$.

Ex. 178. Find the subnormal to any point of the common parabola.

The equation is $\quad y^2 = 2px$.
By differentiating, we obtain

$$\frac{dy}{dx} = \frac{p}{y};$$

$$\therefore y\frac{dy}{dx} = p.$$

The subnormal is equal to half the parameter.

108. *To determine the length of the tangent to any point of a curve referred to rectangular axes.*

In the right-angled triangle PRT, we have

$$\overline{PT}^2 = \overline{PR}^2 + \overline{TR}^2,$$

$$= y^2 + y^2\frac{dx^2}{dy^2};$$

$$\therefore PT = y\sqrt{1 + \frac{dx^2}{dy^2}}.$$

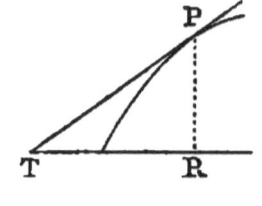

109. *To determine the length of the normal to any point of a curve referred to rectangular axes.*

In the adjoining figure, PN is the normal, and we are required to find its length.

In the right-angled triangle RPN,

$$\overline{PN}^2 = \overline{PR}^2 + \overline{RN}^2,$$

$$= y^2 + \frac{y^2 dy^2}{dx^2};$$

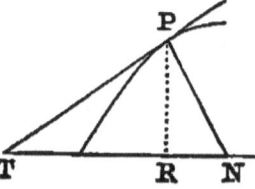

APPLICATION OF THE CALCULUS TO CURVES. 167

$$\therefore \mathrm{PN} = y\sqrt{1+\frac{dy^2}{dx^2}}.$$

We shall now apply these formulas to lines of the second order embraced in the equation

Ex. 179. $\qquad y^2 = mx + nx^2.$ (1)

By differentiating, we obtain

$$\frac{dx}{dy} = \frac{2y}{m+2nx},$$ (2)

and $\dfrac{dy}{dx} = \dfrac{m+2nx}{2y} = \dfrac{m+2nx}{2\sqrt{mx+nx^2}} =$ tan. of the angle which the tangent line makes with the axis of x.

Multiplying both members of the last equation by y, we have

$$y\frac{dy}{dx} = \frac{m+2nx}{2} = \text{the subnormal.}$$

Also, multiplying (2) by y,

$$y\frac{dx}{dy} = \frac{2y^2}{m+2nx} = \frac{2(mx+nx^2)}{m+2nx} = \text{subtangent,}$$

and $y\sqrt{1+\dfrac{dx^2}{dy^2}} = \sqrt{mx+nx^2 + 4\left(\dfrac{mx+nx^2}{m+2nx}\right)^2} = $ tangent;

also $y\sqrt{1+\dfrac{dy^2}{dx^2}} = \sqrt{mx+nx^2 + \tfrac{1}{4}(m+2nx)^2} = $ normal.

If we attribute proper values to m and n, the formulas above will be applicable to each of the conic sections.

In the case of the parabola, $n = 0$; hence we have for the tangent of the angle which a tangent line makes with the axis of x,

$$\frac{dy}{dx} = \frac{m}{2y},$$

$$\frac{y dx}{dy} = 2x = \text{subtangent,}$$

$$y\sqrt{1+\frac{dx^2}{dy^2}} = \sqrt{mx+4x^2} = \text{tangent,}$$

$$y\frac{dy}{dx} = \frac{m}{2} = \text{subnormal,}$$

$$y\sqrt{1+\frac{dy^2}{dx^2}} = \sqrt{mx+\frac{m^2}{4}} = \text{normal.}$$

Ex. 180. In the case of the ellipse, take the equation
$$A^2y^2 + B^2x^2 = A^2B^2. \qquad (1)$$
By differentiating, we have
$$\frac{dy}{dx} = -\frac{B^2x}{A^2y} = \text{the tangent of the angle which a tangent}$$
line makes with the axis of x.

If we place this equal to zero, we shall have
$$x = 0,$$
and from (1), $\qquad y = \pm B,$
which are the co-ordinates of the points at which a tangent line is parallel to the axis of x.

If we put $\frac{dy}{dx}$ equal to *infinity*, the denominator becomes equal to *zero;* that is,
$$y = 0,$$
and $\qquad x = \pm A.$

The tangent line to either of those points is then perpendicular to the axis of x.

Again: $\qquad y\dfrac{dx}{dy} = -\dfrac{A^2y^2}{B^2x},$

$$= -\frac{A^2 - x^2}{x} = \text{subtangent,}$$

and $\qquad y\dfrac{dy}{dx} = -\dfrac{B^2x}{A^2} = \text{subnormal.}$

110. We perceive that the value of the subtangent on the axis of x does not comprise B, and therefore it will be the same whatever be the length of the minor axis; consequently, if any other ellipse, or a circle, be described on the same major axis, the tangent at the point in which the ordinate to the point of tangency meets this new ellipse or circle will cut the axis of x in the same point as the tangent to the first ellipse.

111. In the circle, A becomes equal to B; hence
$$\text{the subtangent} = -\frac{y^2}{x},$$
$$\text{the tangent} = \frac{Ry}{x},$$
$$\text{the subnormal} = -x,$$
$$\text{the normal} = R.$$

APPLICATION OF THE CALCULUS TO CURVES. 169

112. *To determine the* equations *of tangent and normal lines to plane curves.*

The equation of a straight line passing through a given point is (Analytics, Art. 10)
$$y-y'=a(x-x'), \qquad (1)$$
where a denotes the tangent of the angle which the line makes with the axis of x when referred to rectangular co-ordinates.

We have found that the first differential coefficient of y regarded as a function of x represents the tangent of that angle, Art. 102.

Or, $\qquad \dfrac{dy}{dx}=a.$

Hence the equation of a tangent line to a curve at any point is
$$y-y'=\frac{dy}{dx}(x-x').$$

And since the normal is perpendicular to the tangent at the point of tangency, we shall have
$$y-y'=-\frac{dx}{dy}(x-x')$$
for *the equation of the normal.*

To find the equation of a tangent line to any curve referred to rectangular axes, we have the following

RULE.

Differentiate the equation of the curve, determine the value of $\dfrac{dy}{dx}$, *and substitute that value in the equation*
$$y-y'=\frac{dy}{dx}(x-x').$$

Examples.

Ex. 181. Determine the equation of a tangent line to the ellipse at the point whose co-ordinates are x' and y'.
The equation of the ellipse referred to its centre is
$$A^2y^2+B^2x^2=A^2B^2.$$
By differentiation, we have
$$\frac{dy}{dx}=-\frac{B^2x}{A^2y};$$
H

$$\therefore y - y' = -\frac{B^2 x}{A^2 y}(x - x'),$$

which, by reduction, becomes
$$A^2 yy' + B^2 xx' = A^2 B^2,$$
the equation required;

and
$$y - y' = \frac{A^2 y}{B^2 x}(x - x'),$$
the equation of the normal.

Ex. 182. Determine the equation of a tangent line to the circle at a point whose co-ordinates are x', y'.

The equation of the circle is
$$y^2 + x^2 = R^2, \qquad (1)$$
and, by differentiating, we have
$$\frac{dy}{dx} = -\frac{x}{y}.$$

Substituting this in the equation
$$y - y' = \frac{dy}{dx}(x - x'),$$

we have
$$y - y' = -\frac{x}{y}(x - x'),$$
which, by reduction, becomes
$$yy' + xx' = R^2$$
for *the equation of the tangent line to the circle.*

Ex. 183. Find the equation of a tangent line to the parabola at a point whose co-ordinates are x', y'.

The equation of the parabola is
$$y^2 = 2px.$$
And, by differentiating, we have
$$\frac{dy}{dx} = \frac{p}{y}.$$

Substitute this in the equation
$$y - y' = \frac{dy}{dx}(x - x'),$$

and we have
$$y - y' = \frac{p}{y}(x - x'),$$
or
$$y^2 - yy' = px - px'.$$
But
$$y^2 = 2px;$$
$$\therefore 2px - yy' = px - px',$$
$$\therefore yy' = p(x + x'),$$
the *equation of the tangent line to the parabola.*

115. *We may also free the terms of an equation from their exponents.*

Ex. 184. Let $\quad P^n = Q,\quad$ (1)
where P and Q are functions of x.
Then, by differentiating, we have
$$nP^{n-1}dP = dQ;\quad (2)$$
and multiplying both members by P, we obtain
$$nP^n dP = PdQ;$$
and substituting for P^n its value,
then $\quad nQdP = PdQ.\quad$ (3)

Ex. 185. Eliminate the exponents from the equation
$$xy = ae^x + be^{-x}.$$
By differentiating, we obtain
$$xdy + ydx = ae^x dx - be^{-x} dx.$$
Again: by differentiating, we have
$$dxdy + xd^2y + dxdy = ae^x dx^2 + be^{-x} dx^2;$$
or, by reducing and dividing by dx^2,
$$2\frac{dy}{dx} + x\frac{d^2y}{dx^2} = ae^x + be^{-x} = xy;$$
$$\therefore x \cdot \frac{d^2y}{dx^2} + 2\frac{dy}{dx} - xy = 0,$$
which is freed from the exponents.

ASYMPTOTES.

116. *An asymptote* is a line toward which the curve is continually approaching, and which becomes a tangent to the curve at an infinite distance from the origin of co-ordinates.

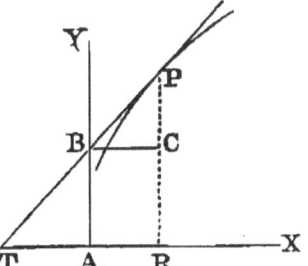

Let the curve be referred to the rectangular axes AX, AY, and suppose P to be the point of tangency at an infinite distance from the origin A. From the equation of the curve find the value of the subtangent TR, which is $y\frac{dx}{dy}$; take the difference between this subtangent and the abscissa x, which is AR, and find the limit of this value when $x = \infty$. If this is finite, lay off that distance equal to AT.

Now in the two similar triangles, TRP and TAB, we have $\frac{TR}{RP} = \frac{TA}{AB}$; then take the limit of the value of RP when $x = \infty$, and determine the value of AB from the above equation.

If AT be finite and AB infinite, the asymptote is *perpendicular* to the axis of x.

If AB is imaginary and AT finite, the curve has no asymptote.

If AT is infinite and AB finite, the asymptote is *parallel* to the axis of x.

If AT and AB both become zero when x is infinite, the asymptote will pass through the origin of co-ordinates; and since only one point of the line is thus determined, its direction must be found by seeking the value of $\frac{dy}{dx}$.

Ex. 186. Let $y = \frac{b}{a}\sqrt{2ax + x^2}$, the equation of the hyperbola.

$$\therefore \frac{dy}{dx} = \frac{b}{a} \cdot \frac{a+x}{\sqrt{2ax+x^2}};$$

$$y\frac{dx}{dy} = \frac{(2a+x)x}{a+x} = \text{subtan.} = TR;$$

$$TR - AR = \frac{2ax+x^2}{a+x} - x = \frac{ax}{a+x} = AT.$$

The limit of the value of AT when $x = \infty$ is a (Art. 10); the limit of the value of the subtangent when $x = \infty$ is x; and the limit of the value of RP or y is $\frac{b}{a}x$.

$$\therefore \frac{x}{\frac{bx}{a}} = \frac{a}{AB};$$

$$\therefore AB = b, \text{ and } AT = a.$$

The line drawn through these two points will be the asymptote.

If we had taken the equation of the hyperbola referred to its centre, $y = \frac{b}{a}(x^2 - a^2)^{\frac{1}{2}}$,

then
$$y\frac{dx}{dy} = \frac{x^2 - a^2}{x},$$

and
$$\frac{x^2 - a^2}{x} - x = -\frac{a^2}{x}.$$

Now, when $x = \infty$, this expression becomes equal to zero; hence the asymptote passes through the origin.

117. The *equations* of the asymptotes of curves may be found independently of the Differential Calculus. Thus, in the hyperbola:

Ex. 187. Let $\quad y = \frac{b}{a}(x^2 - a^2)^{\frac{1}{2}}.$

Expanding by the Binomial Theorem,
$$y = \frac{b}{a}(x - \frac{a^2}{2x} - \frac{a^4}{8x^3} -, \text{etc.}).$$

If we now make $x = \infty$, we shall have
$$y = \frac{b}{a}x,$$
which is the equation to the infinite part of the curve, and consequently to the rectilineal asymptote, since it is only of one dimension. It is also the equation of a line passing through the origin of co-ordinates.

Ex. 188. If we were to take
$$y = \frac{b}{a}\sqrt{2ax + x^2},$$
and expand by the Binomial Theorem, we would have
$$y = \frac{b}{a}\left[x + a - \frac{a^2}{2x} + \frac{a^3}{2x^2} -, \text{etc.}\right].$$

Making $x = \infty$, we shall have
$$y = \frac{b}{a}x + b,$$
which is *the equation to the asymptote.*

Ex. 189. Again: let
$$x^3 - 3axy + y^3 = 0$$
be the equation of the curve; and suppose that
$$y = a'x + b'$$
be the equation of a rectilineal asymptote. Then, by substitution, we shall have
$$(a'^3 + 1)x^3 + 3a'(a'b' - a)x^2 +, \text{etc.} = 0.$$

Whence, by the theory of indeterminate coefficients,

$$a'^3+1=0; \therefore a'=-1.$$
$$a'b'-a=0; \therefore b'=-a.$$

And *the equation* of the asymptote is
$$y=-x-a.$$

118. The method we have just taken has enabled us to determine the equation of the *rectilineal* asymptote of a curve; it is obvious that it will enable us to obtain the equations to curvilineal asymptotes also.

For if we solve the equation of a curve with respect to y, and find it of the form

$$y=ax^n+bx^{n-1}+, \text{ etc.}+h+\frac{k}{x}+, \text{ etc.,}$$

and then make $x=\infty$, the equation of the curve at an infinite distance becomes

$$y=ax^n+bx^{n-1}+, \text{ etc.}+h,$$

which is also the equation of an asymptotic curve of n dimensions.

If $n=1$, then $y=ax+b$, which is also the equation of a rectilineal asymptote, Art. 112.

If $n=2$, then $y=ax^2+bx+c$, which is the equation of a common parabolic asymptote.

Ex. 190. Let the equation of the curve be
$$x^3=(x-a)y^2;$$
then
$$y^2=\frac{x^3}{x-a},$$
or
$$y=\pm\frac{x^{\frac{3}{2}}}{(x-a)^{\frac{1}{2}}}.$$

As x approaches the value of a, both y and $\frac{dy}{dx}$ increase also without limit, and
$$x=a$$
is the equation of a rectilineal asymptote.

If we put y in the form
$$y=\pm x\left(1-\frac{a}{x}\right)^{-\frac{1}{2}},$$
and expand by the Binomial Theorem, we have
$$y=\pm x\left[1+\frac{a}{2x}+\frac{3a^2}{8x^2}+, \text{ etc.}\right]. \tag{1}$$

Hence $y = \pm(x + \tfrac{1}{2}a)$
are the equations of two rectilineal asymptotes.
Again: from (1) we have
$$y = \pm\left(x + \frac{a}{2} + \frac{3a^2}{8x}\right),$$
which are *the equations of two asymptotic curves* of the second order.

Ex. 191. Find the equation of the asymptote to the curve whose equation is
$$y^3 = x^2(2a - x).$$
Ans. $y = -x + \tfrac{2}{3}a$.

Ex. 192. Find the equations of the asymptote to the curve $\quad y^2(x - 2a) = x^3 - a^3$.
Ans. $x = 2a$,
and $y = \pm(x + a)$.

Ex. 193. Find the equation of the asymptote to the curve $\quad y^2(ay + bx) = a^2y^2 + b^2x^2$.
Ans. $y = -\dfrac{b}{a}x + 2a$.

Ex. 194. Find the equation of the asymptote to the curve $\quad x^2y^2 = a^2(x^2 - y^2)$.
Ans. $y = \pm a$.

SINGULAR POINTS OF CURVES.

119. In a curve whose nature is expressed in a general manner by the equation
$$y = f(x),$$
or $\quad f(x, y) = 0$,
it is evident that if we attribute different values to the principal variable, we shall generally obtain different values for the function and its differential coefficients; and when any of these functions attains a value attended with some peculiarity either in its value or its form, the point in the curve corresponding to this will be distinguished by something peculiar in its character. Points of this description are usually denominated *Singular Points*, the principal of which are characterized by the circumstances explained and exemplified in the following articles:

120. *To determine the position of a curve at any point with respect to its tangent.*

Let PP'P'' be any curve, and let x and y be the co-ordinates of the point P. Give to x the several increments, RR', R'R'', each equal to h, and draw the ordinates R'P', R''P''.
Join PP', and produce it to B; also join P'P''; draw PD and P'D' parallel to AR''. Then

$$P'R' = f(x+h) = y + \frac{dy}{dx}h + \frac{d^2y}{dx^2}\frac{h^2}{1.2} +, \text{etc.,} \quad (1)$$

and $$P''R'' = y + \frac{dy}{dx}\frac{2h}{1} + \frac{d^2y}{dx^2}\frac{4h^2}{2} +, \text{etc.}; \quad (2)$$

$$\therefore P'R' - PR = P'D = \frac{dy}{dx}h + \frac{d^2y}{dx^2}\frac{h^2}{2} +, \text{etc.} \quad (3)$$

Also, $$P''R'' - P'R' = P''D' = \frac{dy}{dx}h + \frac{d^2y}{dx^2}\frac{3h^2}{2} +, \text{etc.}; \quad (4)$$

$$\therefore P''D' - P'D = P''B = \frac{d^2y}{dx^2}h^2 +, \text{etc.} \quad (5)$$

Now it is manifest that if the curve lie above its tangent at P', or be convex to the axis of x, P''D' is greater than BD', and therefore we must have the *second differential coefficient positive*.

But if the curve lie below its tangent at P', or be concave to the axis of x, D'P'' would evidently be less than BD', and therefore *the second differential coefficient would be negative*.

Also, since by the continual diminution of h the first term $\frac{d^2y}{dx^2}.h^2$ will become greater than the sum of all the succeeding terms of the series, it follows that the algebraic sign of this series will be the same as that of its first term, and therefore, according as a curve is convex or concave to the axis of abscissas, the second differential coefficient $\left(\frac{d^2y}{dx^2}\right)$ will be positive or negative.

Hence, conversely, a curve will be convex or concave to the axis of abscissas according as the second differential coefficient of the ordinate is positive or negative. Or, more generally,

A curve will be convex or concave to the axis of abscissas according as the ordinate and its second differential coefficient have the same or different algebraic signs.

121. A singular point of a curve is a point which has some peculiarity which the other points do not generally have; as, for instance,

1. The point at which a tangent line is either parallel or perpendicular to either axis.

2. The point at which the curve changes from being convex to concave to the axis of x, or from being concave to convex. The point at which this change of curvature takes place is called a *point of inflection*.

3. *A multiple point;* that is, a point at which two or more branches of a curve intersect each other.

4. *A cusp;* that is, a point at which two or more branches of a curve terminate and have a common tangent. If the branches lie on different sides of the tangent, it is called a cusp of the *first* order; if on the same side, a cusp of the *second order.*

5. *An isolated point;* that is, a point whose co-ordinates satisfy the equation of the curve, although the point is entirely detached from every other point of the curve.

The first of these has been discussed in Art. 103.

122. In order to determine whether a proposed curve is convex or concave toward the axis of x, *differentiate the equation of the curve twice; then, if the ordinate and second differential coefficient have the same sign (whatever be the value attributed to the variable), the curve is convex; if they have different signs, the curve is concave* (Art. 120).

Ex. 195. As a first example, let us take the equation of a circle referred to the centre as the origin of rectangular axes, $y^2 = R^2 - x^2$.

Differentiating, we have $\dfrac{dy}{dx} = -\dfrac{x}{y}$,

and $\dfrac{d^2y}{dx^2} = -\dfrac{x^2 + y^2}{y^3} = -\dfrac{R^2}{y^3}$.

When y is positive, the second differential coefficient is negative; and when y is negative, the second differential coefficient is positive. Hence the curve is entirely concave toward the axis of x.

Ex. 196. Again: let us take the general equation of the circle
$$(y-\beta)^2+(x-\alpha)^2=R^2;$$
$$\therefore y=\beta\pm\sqrt{R^2-(x-\alpha)^2}.$$
And differentiating,
$$\frac{dy}{dx}=\mp\frac{x-\alpha}{\sqrt{R^2-(x-\alpha)^2}},$$
and
$$\frac{d^2y}{dx^2}=\mp\frac{R^2}{\{R^2-(x-\alpha)^2\}^{\frac{3}{2}}}.$$

If the upper sign be used in the value of y, the corresponding part of the curve is concave toward the axis of x; but if the lower sign be used, it is convex.

Ex. 197. Let the curve be the common parabola
$$y^2=2px.$$

Ex. 198. Determine whether the ellipse is concave or convex toward the axis of x.

123. To determine whether a proposed curve has a point of inflection.

Differentiate the equation of the curve twice; then, if the second differential coefficient changes its sign (by attributing different values to the variable), the point at which this change takes place is a point of inflection. Therefore, Art. 29, there must be one value of the second differential coefficient equal to zero or infinity, and the roots of the equations
$$\frac{d^2y}{dx^2}=0, \text{ or } \frac{d^2y}{dx^2}=\infty,$$
will give the abscissas of the point of inflection.

Ex. 199. Determine whether the curve whose equation is $\quad y=a+c(x-b)^3$
has a point of inflection.

Differentiating, we have
$$\frac{dy}{dx}=3c(x-b)^2,$$
$$\frac{d^2y}{dx^2}=6c(x-b).$$

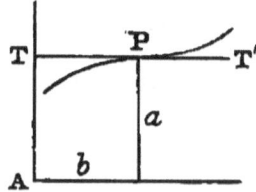

SINGULAR POINTS OF CURVES. 179

When $x=b$, the first differential coefficient is zero, and the tangent is parallel to the axis of x at the point whose abscissa is b. Substituting this value of x in the equation of the curve, we get $y=a$.

Examining the second differential coefficient, we find that when x is less than b, the second differential coefficient is negative, and the curve is concave up to that point; when $x>b$, the second differential coefficient is positive, and the curve is convex; hence there is an inflection of the curve at the point whose co-ordinates are
$$x=b,\ y=a.$$

Ex. 200. Determine whether the curve whose equation is
$$y=b-c(x-a)^3,$$
has a point of inflection.

Ex. 201. Determine whether the curve whose equation is
$$y=x+36x^2+2x^3-x^4$$
has a point of inflection.

It has two points of inflection.

124. To determine whether a proposed curve has a multiple point.

If, when a certain value is assigned to the abscissa x, we find only one value for the ordinate y, but more than one value for $\dfrac{dy}{dx}$, it is evident that the curve has two or more branches intersecting at that point; for $\dfrac{dy}{dx}$ expresses the tangent of the angle which a tangent line to the curve makes with the axis of x; hence there must be more than one rectilineal tangent at the corresponding point, which is therefore a multiple point, and its degree of multiplicity is expressed by the number of real values of $\dfrac{dy}{dx}$.

Ex. 202. Let the equation of the curve proposed be
$$y=\pm x(a^2-x^2)^{\frac{1}{2}}.\quad (1)$$
Differentiating, we have
$$\frac{dy}{dx}=\pm\frac{a^2-2x^2}{(a^2-x^2)^{\frac{1}{2}}}.$$
For every value of x in (1) we have two values of

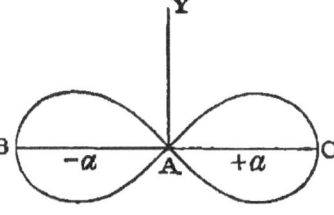

y numerically equal with opposite signs; hence the curve has two branches symmetrical with regard to the axis of x.

Also, when $x=\pm a$, $y=0$; that is, the curve cuts the axis of x at a distance equal to a to the right, and a to the left of the origin.

When $x=0$, $y=0$; hence the two branches of the curve intersect at the origin of co-ordinates. The origin is therefore a *multiple point*.

Ex. 203. Let the equation of the curve proposed be
$$y=b\pm(x-a)\sqrt{x}.$$

125. To ascertain whether a proposed curve has a cusp.

If P be a point of a curve at which PR is a tangent, then, if the ordinates P'R' and P"R", drawn to the consecutive points on each side of the point P, be both *less* or both *greater* than the ordinate at P, it is evident that P will be the point at which the two branches of the curve will unite; or, in other words, be a *cusp of the first order*.

Ex. 204. Let $(y-a)^3-(x-b)^2=0$, be the equation of a curve. It is required to determine whether it has a cusp.

Transposing and extracting the root, we have
$$y-a=(x-b)^{\frac{2}{3}};$$
$$\therefore \frac{dy}{dx}=\frac{2}{3(x-b)^{\frac{1}{3}}},$$
and
$$\frac{d^2y}{dx^2}=-\frac{2}{9(x-b)^{\frac{4}{3}}}.$$

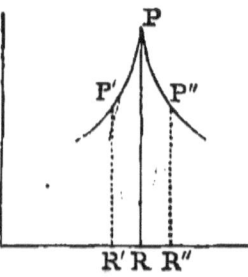

If we make $x=b$, the first differential coefficient will become infinite, and the tangent will be perpendicular to the axis of x at the point whose co-ordinates are $x=b$, $y=a$.

In regard to the second differential coefficient, it will be *negative* for every value of x, since the factor $(x-b)$ is raised to an even power. Hence the curve will be concave toward the axis of x.

SINGULAR POINTS OF CURVES. 181

If we make $x=b+h$ or $x=b-h$, in either case we shall have
$$y=a+h^{\frac{2}{3}},$$
and therefore y will be less for $x=b$ than for any other value of x, either greater or less than b. Hence the value $x=b$ is the abscissa of the least ordinate; and there is a cusp of the first order at the point whose coordinates are $x=b$, $y=a$.

Ex. 205. Again: let $(y-a)^3+(x-b)^2=0$.

Then $\quad y-a=-(x-b)^{\frac{2}{3}},$

and $\quad \dfrac{dy}{dx}=-\dfrac{2}{3(x-b)^{\frac{1}{3}}},$

$\dfrac{d^2y}{dx^2}=\dfrac{2}{9(x-b)^{\frac{4}{3}}}.$

As before, $x=b$ makes the first differential coefficient equal to infinity.

The second differential coefficient will be *positive* for every value of x either greater or less than b. The curve will therefore be convex toward the axis of x.

If we make $x=b+h$, or $x=b-h$, we shall have, in either case,
$$y=a-h^{\frac{2}{3}};$$
and hence y will be greater for $x=b$ than for any other value of x either greater or less than b. Hence the value $x=b$ is the abscissa of the greatest ordinate. The branches PP' and PP" are not considered as parts of a continuous curve, for the general relations between x and y which determine each of those parts are entirely broken at the point P, where $x=b$.

Ex. 206. Let $\quad (y-a)^2=(x-b)^3;\quad$ (1)

then $\quad y=a\pm(x-b)^{\frac{3}{2}},$

and $\quad \dfrac{dy}{dx}=\pm\tfrac{3}{2}(x-b)^{\frac{1}{2}},$

$\dfrac{d^2y}{dx^2}=\pm\dfrac{3}{4(x-b)^{\frac{1}{2}}}.$

If we put the first differential coefficient equal to zero,

we shall have $x=b$, and substituting this value of x in the equation of the curve, we find $y=a$. Hence the point whose co-ordinates are $x=b$, $y=a$, is a point of the curve at which a tangent line is parallel to the axis of x.

Now for $x=b-h$ the ordinates are imaginary, and for $x=b+h$ the values of
$$\frac{d^2y}{dx^2} = \pm \frac{3}{4\sqrt{h}}.$$

Hence there is a cusp at P, where the upper branch, PP', which is convex toward the axis of x, is determined by $+\dfrac{3}{4\sqrt{h}}$,

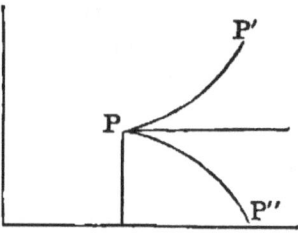

and the lower branch, PP'', is concave, determined by
$$-\frac{3}{4\sqrt{h}}.$$

If for x in (Eq. 1) we substitute $b+h$, we shall have
$$y = a \pm h^{\frac{3}{2}};$$
and giving to h a succession of values, which shall have small differences, we may trace the curve.

Ex. 207. Let $(y-x^2)^2 = x^5$ be the equation of the curve; then
$$y = x^2 \pm x^{\frac{5}{2}}, \qquad (2)$$
$$\frac{dy}{dx} = 2x \pm \tfrac{5}{2} x^{\frac{3}{2}},$$
and
$$\frac{d^2y}{dx^2} = 2 \pm \tfrac{15}{4} x^{\frac{1}{2}}.$$

If we make $x=0$, $y=0$; hence both branches of the curve pass through the origin. Putting the first differential coefficient equal to *zero*, we have
$$2x \pm \tfrac{5}{2} x^{\frac{3}{2}} = 0.$$

One value of x is *zero;* this shows that the axis of x is a tangent to the curve at the origin. Examining the second differential coefficient, we see that the upper branch, corresponding to $2+\tfrac{15}{4}x^{\frac{1}{2}}$, is continually convex toward the axis of x, while the lower branch is convex from the origin to the point whose abscissa is $\tfrac{64}{225}$, and concave beyond that point. If $x=1$, then $y=0$, and the lower branch cuts the axis of x at that point.

SINGULAR POINTS OF CURVES. 183

Ex. 208. Let $(x-y^2)^2 = y^5$ be the equation.
Then $x = y^2 \pm y^{\frac{5}{2}}$;

$$\therefore \frac{dy}{dx} = \frac{1}{2y \pm \frac{5}{2}y^{\frac{3}{2}}}.$$

Putting $\dfrac{dy}{dx} = \infty$,

we have $2y \pm \frac{5}{2}y^{\frac{3}{2}} = 0$;
$y = 0$, and $x = 0$.

Hence the axis of y is tangent to both branches of the curve at the origin of co-ordinates. If y be made negative, x becomes imaginary, which shows that the curve does not pass below the axis of x.

In the equation

$$x = y^2 \pm y^{\frac{5}{2}},$$

if y be less than unity, $y^{\frac{5}{2}}$ will be less than y^2, and x will have two positive values, represented by RP and RP'. Therefore the point A is a cusp of the second order.

This example and the one preceding it will give the student an idea of *cusps of the second order*.

126. *To determine whether a proposed curve has an isolated point.*

If, when a certain value is given to the abscissa x, the value of the ordinate y remains possible at the same time that one or more of its differential coefficients become imaginary, the corresponding point of the curve is called an *isolated* or *conjugate point*. Hence, to determine the positions of such points, we must ascertain what possible values of the co-ordinates satisfy the equation of the curve at the same time that they render one or more of the differential coefficients imaginary.

Ex. 209. Let us take the curve whose equation is

$$ay^2 = x(x+b)^2;\qquad(1)$$

$$\therefore y = \pm(x+b)\sqrt{\frac{x}{a}},\qquad(2)$$

$$\frac{dy}{dx} = \pm \frac{1}{2\sqrt{a}}\left[3\sqrt{x} + \frac{b}{\sqrt{x}}\right].\qquad(3)$$

Now if $x=-b$ in (1), we shall have $y=0$, and the corresponding value of
$$\frac{dy}{dx} = \mp\sqrt{-\frac{b}{a}}$$
will cause the values preceding and succeeding this to be imaginary; so that, if A be the origin of the co-ordinates, and AP be taken $=b$, the point P belongs to the curve, but it is entirely detached from the rest of the figure, making the point an *isolated point*.

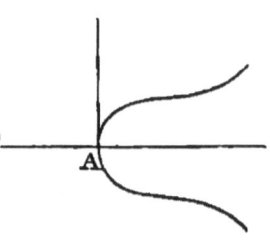

Ex. 210. Let $\quad ay^2 = x^3 - bx^2$
be the equation of the proposed curve. Then, if $x=0$, $y=0$. The co-ordinates of the origin satisfy the equation, and we might at first suppose that the curve passes through that point.

Solving the equation for y, we have
$$y = \pm x\left(\frac{x-b}{a}\right)^{\frac{1}{2}}.$$

Now for all values of x less than b, y will be imaginary; and when $x=b$, $y=0$. Hence the curve cuts the axis of x at a distance equal to b from the origin of co-ordinates. For every value of x greater than b, y will have two values numerically equal with contrary signs.

Differentiating, we have
$$\frac{dy}{dx} = \frac{3x-2b}{2\sqrt{a(x-b)}};$$
and making $x=0$, $\dfrac{dy}{dx} = -\dfrac{2b}{2\sqrt{-ab}}$,
an imaginary expression.

Hence the origin of co-ordinates is an isolated point, and its co-ordinates satisfy the equation of the curve; but there is no point of the curve which is consecutive with it.

Transcendental Curves.

127. Curves may be divided into two general classes, *algebraic* and *transcendental*. When the relation be-

tween the abscissa and ordinate of a curve can be expressed in algebraic terms, the curve is called an *algebraic* curve; but when this relation can not be expressed without the aid of transcendental terms, the curve is then called a *transcendental* curve.

THE LOGARITHMIC CURVE.

128. The *logarithmic* curve takes its name from the property that, when referred to rectangular axes, one of the co-ordinates is equal to the logarithm of the other.

Take the equation $a^x = y$;
then (Alg., Art. 153) x is equal to the logarithm of y, taken in a system whose base is a.

In the diagram, let A be the origin of rectangular axes. Then,
if $x=0$, $y=1=AB$;
if $x=1$, $y=a$;
if $x=2$, $y=a^2$;
  etc., etc.

If $x=-1$, $y=a^{-1}=\dfrac{1}{a}$.

If $x=-2$, $y=a^{-2}=\dfrac{1}{a^2}$.

  etc., etc.

If a, the base of the system, is greater than unity, the logarithms of all numbers which are less than unity will be negative. Hence these are represented by the parts of the line AD to the left of the origin.

When $y=0$, x must be infinite and negative, as
$$a^{-x}=0,$$
or
$$\dfrac{1}{a^x}=0.$$

Taking the equation
$$x = \log. y, \qquad (1)$$
and differentiating, putting m = modulus of the system, we shall obtain
$$\dfrac{dx}{dy} = \dfrac{m}{y}, \qquad (2)$$
or
$$y\dfrac{dx}{dy} = m. \qquad (3)$$

Equation (2) may be expressed
$$\frac{dy}{dx} = \frac{y}{m}; \qquad (4)$$
and as this represents the tangent of the angle which a tangent line makes with the axis of x, therefore the tangent will be parallel to the axis of x when $y=0$; but when $y=0$, $x=\infty$. The axis of x is therefore an asymptote to the curve.

129. (Eq. 3) represents the subtangent, *which is equal to the modulus of the system of logarithms from which the curve is constructed.*

In the Naperian system, $m=1$, and the subtangent will be equal to AB.

The Cycloid.

130. *A cycloid is a curve which is described by a point in the circumference of a circle while the circle is moving along a straight line.*

The circle is called the *generating circle*, and the point which describes the cycloid is called the *generating point*.

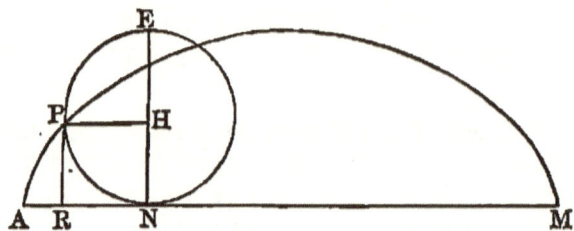

Let EPN be the generating circle, A the origin of co-ordinates. Let us suppose that the point P coincided with the point A, and that it has described the arc AP while the circle was rolling along the straight line AM. Then, if

AR=x, and RP=y=NH,

and r=radius of the generating circle,

we shall have

PH=$\sqrt{2ry-y^2}$=RN. (Geom., B. IV., Theor. IX.)

Now AR=AN−RN; but AN=arc PN, and HN is the versine of the arc PN.

$$\therefore x = \text{versin.}^{-1} y - \sqrt{2ry - y^2}, \quad (1)$$

which is the *transcendental equation of the cycloid*.

By differentiating (1), we obtain

$$dx = \frac{y\,dy}{\sqrt{2ry - y^2}}, \quad (2)$$

which is the *differential equation of the cycloid*.

131. To find the normal, we have

$$y\sqrt{1 + \frac{dy^2}{dx^2}} = y\sqrt{1 + \frac{2r-y}{y}},$$

$$= \sqrt{2ry} = normal.$$

Subnormal $\quad y\dfrac{dy}{dx} = \sqrt{2ry - y^2},$

tangent $\quad y\sqrt{1 + \dfrac{dx^2}{dy^2}} = \dfrac{y\sqrt{2ry}}{\sqrt{2ry - y^2}},$

subtangent $\quad y\dfrac{dx}{dy} = \dfrac{y^2}{\sqrt{2ry - y^2}}.$

If we differentiate the equation

$$dx = \frac{y\,dy}{\sqrt{2ry - y^2}},$$

regarding dx as constant, we obtain, after a little reduction,

$$d^2y = -\frac{r\,dy^2}{2ry - y^2}.$$

Hence the cycloid is concave toward the axis of x.

SPIRALS.

132. A *spiral* is a curve generated by a point which moves along a straight line according to some prescribed law, the straight line having at the same time a uniform angular motion.

The fixed point P, about which the straight line revolves, is called the *pole* of the spiral. That part of the spiral which is formed while the moving line makes one revolution, is called a *spire;* and the distance, PM or PA, which

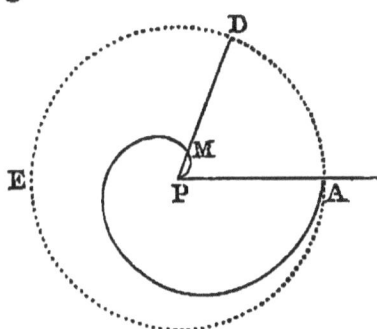

the point has traveled while the line has been in motion, is called the *radius vector*.

133. From the manner in which the spiral is generated, its equation will be
$$r = f(\theta),$$
where r is the radius vector, and θ the arc of the measuring circle.

Spiral of Archimedes.

134. If, while the line PD revolves uniformly about P, a point also moves uniformly along the line PD, that point will describe a spiral, which is the *spiral of Archimedes*.

From the definition, the *radii vectores* are proportional to the measuring arcs estimated from A. That is,
$$PM : PA :: \text{arc } AD : \text{cir. } ADE.$$
Calling PM the radius vector $= r$,
" PA the radius of the circle $ADE = R$,
and θ the measuring arc estimated from A, we shall have
$$r : R :: \theta : 2\pi R;$$
$$\therefore r = \frac{\theta}{2\pi}.$$
If we put
$$\frac{1}{2\pi} = a,$$
then
$$r = a\theta,$$
the equation of the spiral of Archimedes.

Logarithmic Spiral.

135. If, while the straight line is revolving about the pole, a point is also moving along this line in such a manner that the logarithm of the radius vector is constantly proportional to the measuring arcs, the point will describe a spiral which is called the *logarithmic spiral*.

Hence, by the definition,
$$\theta = \log. r$$
is the equation of the logarithmic spiral.

Hyperbolic or Reciprocal Spiral.

136. If, while the straight line is revolving about the pole, a point is moving along this line in such a manner that the *radii vectores* shall be *inversely proportional* to

THE CISSOID.

the corresponding measuring arcs, the point will describe a spiral which is called the *hyperbolic* or *reciprocal spiral*.

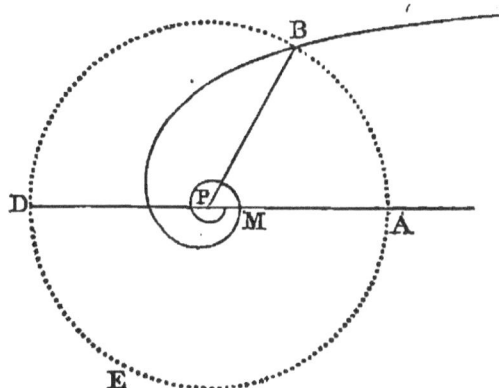

From the definition, we have
$$PM : PB :: \text{arc } AB : \text{cir. } ABE.$$
Calling PM unity, we shall have
$$1 : PB :: \text{arc } AB : \text{cir. } ABE,$$
or
$$1 : r :: \theta :: 2\pi;$$
$$\therefore r = \frac{2\pi}{\theta} = 2\pi\theta^{-1}.$$
Putting $2\pi = a$, we have
$$r = a\theta^{-1} = \frac{a}{\theta},$$
or
$$r\theta = a,$$
which is the equation of the reciprocal spiral.

This spiral is called the *hyperbolic spiral* from the analogy which its equation bears to that of the hyperbola when referred to its asymptotes.

THE CISSOID.

137. If AQB is a semicircle, and we take any equal parts, CM and CN, and draw the ordinates, MR, NQ; join AQ, cutting MR in P; then the *locus* of P will be the curve called the *Cissoid*.

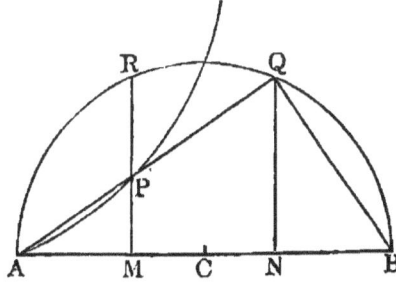

Let A be the origin of rectangular axes.
Put $AB = a =$ the diameter of the circle,
$$x = AM,\ y = MP.$$
Then $\overline{MP}^2 : \overline{AM}^2 :: \overline{QN}^2 : \overline{AN}^2,$
or $y^2 : x^2 :: AN.NB : \overline{AN}^2$; $\because \overline{QN}^2 = AN.NB,$
$$:: NB : AN,$$
$$:: x : a-x;\ \therefore NB = AM;$$
$$\therefore y^2 = \frac{x^3}{a-x}, \tag{1}$$
the equation of the cissoid.

Differentiating (1), we have
$$y\frac{dx}{dy} = \frac{2(ax - x^2)}{3a - 2x}, \text{ for the } subtangent.$$

To find the polar equation of the cissoid:
Take A as the pole, and join QB.
Put $AP = r$, and the angle $QAB = \theta$, and $AB = 2a$;
then $AQ = 2a \cos.\theta,$ and
$$AQ = \frac{QN}{\sin.\theta} = \frac{BN}{\tan.\theta \sin.\theta} = \frac{AM}{\tan.\theta \sin.\theta} = \frac{r \cos.\theta}{\tan.\theta \sin.\theta}. \tag{2}$$

Equating (1) and (2), we have
$$\frac{r . \cos.\theta}{\tan.\theta \sin.\theta} = 2a \cos.\theta;$$
$$\therefore r = \frac{2a \sin.^2 \theta}{\cos.\theta},$$
the polar equation required.

139. *To determine the differential of the length of an arc referred to rectangular axes.*

Let $AR = x$ and $RP = y$ be the co-ordinates of the point P; AP the arc $= z$.

Then, if we give to x an increment, $RR' = h$, $AP' = z'$, we shall have

$$\frac{z' - z}{h} = \frac{\text{arc } PP'}{PD}.$$

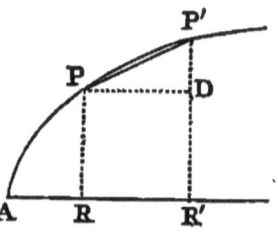

But, by (Art. 11), the arc and chord are ultimately in a ratio of equality. Passing to the limit, we have
$$\frac{dz}{dx} = \text{limit of } \frac{\text{chd. } PP'}{PD},$$

LENGTHS AND AREAS. 191

$$= \text{limit of } \sqrt{\frac{PD^2 + DP'^2}{PD^2}},$$

$$= \text{limit of } \sqrt{1 + \left(\frac{y'-y}{x'-x}\right)^2},$$

$$= \text{limit of } \sqrt{1 + \left(\frac{y'-y}{h}\right)^2},$$

$$= \sqrt{1 + \frac{dy^2}{dx^2}};$$

$$\therefore dz = \sqrt{dx^2 + dy^2},$$

which is the differential of the length of an arc.

140. *To find the differential of the area of a segment of any curve referred to rectangular axes.*

Let ARP be a segment of any curve. Put AR=x, RP=y, the co-ordinates of the point P. Give to x an increment, h=RR'. Then, if we call the area ARP=s, the area of AR'P' will be s', and

$$\frac{s'-s}{h} = \frac{\text{area } RR'P'P}{h}.$$

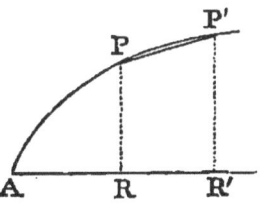

Therefore, if we pass to the limit, we have

$$\frac{ds}{dx} = \text{limit of } \frac{\text{trapezoid } RP'}{RR'},$$

$$= \text{limit of } \frac{(RP + R'P')}{2},$$

$$= \text{limit of } \tfrac{1}{2}(y'+y)$$

$$= y;$$

$$\therefore ds = y\,dx,$$

which is the differential of the area of a segment of any curve referred to rectangular axes.

141. *To find the differential of a surface of revolution.*

If the plane figure AR'P' revolve about the axis AR', and we make the surface generated by the revolution of AP=S, then the surface generated by APP'=S'.

Put AR=x, RP=y, and RR'=h.

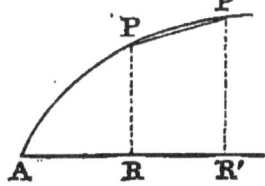

Then $\dfrac{S'-S}{h} = \dfrac{\text{surface generated by arc PP}'}{RR'}$.

If we pass to the limit, we have

$\dfrac{dS}{dx} = \text{limit of } \dfrac{\text{surface generated by chd. PP}'}{RR'}$,

$= \text{limit of } \dfrac{\text{surface of the frustum of cone}}{RR'}$,

$= \text{limit of } \dfrac{\pi(PR+P'R')PP'}{RR'}$ (Mens., Prob. V.),

$= \text{limit of } \pi(y'+y)\sqrt{1+\left(\dfrac{y'-y}{h}\right)^2}$,

$= 2\pi y \sqrt{1+\dfrac{dy^2}{dx^2}}$;

$\therefore dS = 2\pi y \sqrt{dx^2+dy^2}$.

Now $2\pi y$ *is the circumference of a circle perpendicular to the axis, and* $\sqrt{dx^2+dy^2}$ *is the differential of the arc of the generating curve; hence the differential of the surface is equal to the product of those two expressions.*

142. *To find the differential of a solid of revolution.*

If the plane figure AR'P' revolve about the axis AR', it will generate a solid of revolution.

Put $AR=x$, $RP=y$, the co-ordinates of the point P. Give to x an increment, $RR'=h$, and let
V = volume described by AP,
V' = volume described by AP';

then $\dfrac{V'-V}{h} = \dfrac{\text{solid generated by RR'P'P}}{RR'}$.

Passing to the limit, we obtain

$\dfrac{dV}{dx} = \text{limit of } \dfrac{\text{solid generated by RR'P'P}}{RR'}$,

$= \text{limit of } \dfrac{\text{frustum of cone}}{RR'}$,

$= \text{limit of } \dfrac{\pi(\overline{PR}^2 + P'R' \cdot PR + \overline{P'R'}^2)}{3}$,

LENGTHS AND AREAS. 193

$$= \text{limit of } \pi\left(\frac{y^2 + yy' + y'^2}{3}\right),$$
$$= \pi y^2;$$
$$\therefore dV = \pi y^2 dx.$$

Now πy^2 is the area of a circle perpendicular to the axis. Hence *the differential of a volume of revolution is equal to the area of a circle perpendicular to the axis multiplied by the differential of the abscissa of the generating curve.*

The expression $\pi y^2 dx$ is evidently a cylinder whose base is πy^2 and altitude dx.

143. *To determine the differential of the length of a curve referred to polar co-ordinates.*

The differential of the length of a curve referred to rectangular axes is

$$dz = \sqrt{dx^2 + dy^2}.$$

Let PD, the radius vector, be denoted by r, and the angle which it makes with the axis PX be denoted by θ; x and y being the co-ordinates of the point D referred to rectangular axes. Then
$x = r \cdot \cos \theta$,
and $y = r \cdot \sin \theta$,
and $dx = \cos \theta \, dr - r \sin \theta \cdot d\theta$,
and $dy = \sin \theta \, dr + r \cos \theta \cdot d\theta$;
$\therefore dz = \sqrt{\{(\cos \theta \, dr - r \sin \theta \, d\theta)^2 + (\sin \theta \, dr + r \cos \theta \, d\theta)^2\}}$
$$= \sqrt{dr^2 + r^2 d\theta^2},$$

which is *the differential of the arc when referred to polar co-ordinates.*

144. *To find the differential of the area of any segment of a curve referred to polar co-ordinates.*

The area of the triangle PND is equal to $\frac{1}{2}$PN $\times$ ND. $= \frac{1}{2}xy$, and its differential is therefore $= \dfrac{xdy + ydx}{2}$.

Again: the differential of the segment of the curve AND is ydx;
$$\therefore ydx = d(\text{AND}).$$
I

Hence the differential of the area APD must be equal to the difference of these differentials. That is,

diff. of the area $PAD = d(PND) - d(AND)$,

$$= \frac{xdy + ydx}{2} - ydx,$$

$$= \frac{xdy - ydx}{2}.$$

But, by the preceding problem,

$$dy = \sin\theta\, dr + r\cos\theta\, d\theta.$$

Multiplying this by the value of x, which is $r\cos\theta$, we have

$$xdy = r\sin\theta\cos\theta\, dr + r^2\cos^2\theta\, d\theta.$$

And, in like manner,

$$ydx = r\sin\theta\cos\theta\, dr - r^2\sin^2\theta\, d\theta;$$
$$\therefore xdy - ydx = r^2\, d\theta,$$

and $\quad\dfrac{xdy - ydx}{2} = \tfrac{1}{2}r^2\, d\theta = ds,$

which is *the differential of the area of any segment of a curve referred to polar co-ordinates.*

145. If we have an equation to a curve referred to rectangular co-ordinates, we may transform it to one between polar co-ordinates by using the formulas

$$x = r\cos\theta, \qquad (1)$$
$$y = r\sin\theta, \qquad (2)$$

where r is the radius vector, and θ the angle which the radius vector makes with the axis of x.

Differentiating (1) and (2), we obtain

$$\frac{dx}{d\theta} = \cos\theta\frac{dr}{d\theta} - r\sin\theta, \qquad (3)$$

$$\frac{dy}{d\theta} = \sin\theta\frac{dr}{d\theta} + r\cos\theta; \qquad (4)$$

$$\therefore \frac{dy}{dx} = \frac{\dfrac{dy}{d\theta}}{\dfrac{dx}{d\theta}} = \frac{\sin\theta\dfrac{dr}{d\theta} + r\cos\theta}{\cos\theta\dfrac{dr}{d\theta} - r\sin\theta}.$$

which is the tangent of the angle which the tangent line makes with the axis of x. = tan. DCN.

But $\quad <\mathrm{PDC} = \mathrm{DCN} - \mathrm{DPC},$
$$= \mathrm{DCN} - \theta;$$

LENGTHS AND AREAS. 195

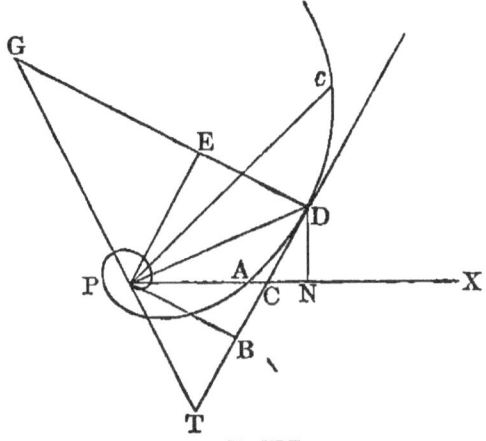

$$\therefore \tan. \text{PDC} = \frac{\tan. \text{DCN} - \tan. \theta}{1 + \tan. \text{DCN} . \tan. \theta},$$

$$= \frac{\dfrac{\sin. \theta \dfrac{dr}{d\theta} + r \cos. \theta}{\cos. \theta \dfrac{dr}{d\theta} - r \sin. \theta} - \tan. \theta}{1 + \left(\dfrac{\sin. \theta \dfrac{dr}{d\theta} + r \cos. \theta}{\cos. \theta \dfrac{dr}{d\theta} - r \sin. \theta}\right) \tan. \theta},$$

$$= \frac{\sin. \theta \dfrac{dr}{d\theta} + r \cos. \theta - \tan. \theta \cos. \theta \dfrac{dr}{d\theta} + r \sin. \theta \tan. \theta}{\cos. \theta \dfrac{dr}{d\theta} - r \sin. \theta + \tan. \theta \sin. \theta \dfrac{dr}{d\theta} + r \cos. \theta \tan. \theta}.$$

But $\sin. \theta \tan. \theta = \dfrac{\sin.^2 \theta}{\cos. \theta}$, and $\cos. \theta \tan. \theta = \sin. \theta$.
Making these substitutions, we have

$$\tan. \text{PDC} = r \frac{d\theta}{dr}. \qquad (5)$$

Now, in the triangle PDT, we have

$$\frac{\text{PT}}{\text{PD}} = \tan. \text{PDT} = r \frac{d\theta}{dr};$$

$$\therefore \text{PT} = \text{PD} \tan. \text{PDT},$$

$$= r^2 \frac{d\theta}{dr} = \text{subtangent}.$$

146. For *the subtangent of any curve referred to polar co-ordinates is a line drawn from the pole perpendicular to a radius vector, and limited by a tangent drawn through the extremity of the radius vector.* Or, in other words, *the subtangent is the projection of the tangent on a line drawn through the pole, perpendicular to the radius vector, which passes through the point of contact.*

147. Since the arc and tangent are coincident at the point of contact, the angle in which the radius vector cuts the arc of the spiral is also the angle PDC. If we denote this angle by ϕ, we shall have

$$\tan \phi = \frac{r d\theta}{dr},$$

or

$$\tan^2 \phi = \frac{r^2 d\theta^2}{dr^2}.$$

But

$$\tan^2 \phi = \frac{\sin^2 \phi}{\cos^2 \phi} = \frac{\sin^2 \phi}{1 - \sin^2 \phi} = \frac{1 - \cos^2 \phi}{\cos^2 \phi};$$

$$\therefore \left(1 + \frac{r^2 d\theta^2}{dr^2}\right) \sin^2 \phi = \frac{r^2 d\theta^2}{dr^2};$$

$$\therefore \sin \phi = \frac{r d\theta}{\sqrt{dr^2 + r^2 d\theta^2}}.$$

In like manner, we may find

$$\cos \phi = \frac{dr}{\sqrt{dr^2 + r^2 d\theta^2}}.$$

148. If we draw PB perpendicular to the tangent, we shall have

$$PB = PD \sin \phi = \frac{r^2 d\theta}{\sqrt{dr^2 + r^2 d\theta^2}},$$

$$DB = PD \cos \phi = \frac{r dr}{\sqrt{dr^2 + r^2 d\theta^2}}.$$

149. It is sometimes convenient to know the relation between the radius vector of a spiral and the perpendicular let fall upon the tangent; hence, if we call the perpendicular $PB = p$, we shall have in all curves referred to polar co-ordinates,

$$p = \frac{r^2}{\sqrt{\frac{dr^2}{d\theta^2} + r^2}}.$$

Also
$$\frac{d\theta}{dr} = \frac{p}{r\sqrt{r^2-p^2}},$$

or
$$\frac{dr}{d\theta} = \frac{r\sqrt{r^2-p^2}}{p}.$$

NORMALS.

150. *To draw a normal to a spiral, and find the length of the subnormal.*

Let DG, perpendicular to the tangent at the point of tangency, meet the line TPG in G, drawn perpendicular to the radius vector PD; then PG is called the *polar subnormal*, and DG the *normal*.

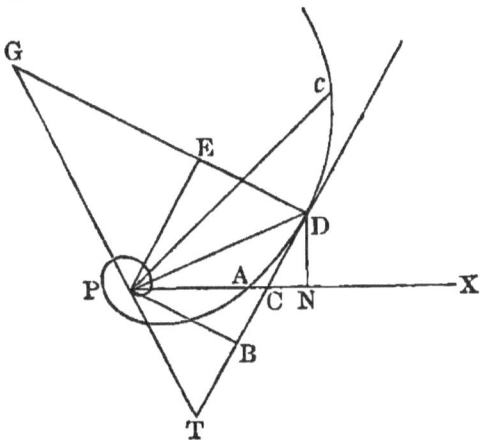

Then in the triangle GPD, we shall have
PG = PD tan. GDP = PD cot. PDT,
$$= r\frac{dr}{rd\theta} = \frac{dr}{d\theta}.$$

And the normal DG = $\sqrt{\overline{PD}^2 + \overline{PG}^2}$,
$$= \sqrt{r^2 + \frac{dr^2}{d\theta^2}}.$$

ASYMPTOTES.

151. *To ascertain whether a spiral admits of an asymptote, and to determine its position.*

In the equation $r = f(\theta)$, find the value of θ in terms of r. Then, if this value and that of the subtangent, which is $\dfrac{r^2 d\theta}{dr}$, do not become infinite when the radius vector is indefinitely increased, the spiral has an asymptote, and its position may be determined.

For, if PR be the direction of the radius vector when infinite, and the subtangent PT be drawn perpendicular to it, then TS, parallel to PR, will evidently be the asymptote.

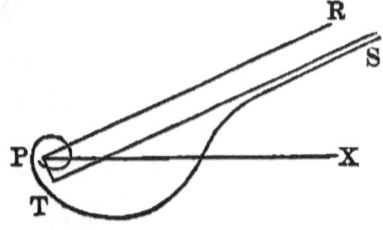

Ex. 211. Let us take the hyperbolic spiral
$$r = \frac{a}{\theta};$$
$$\therefore \theta = \frac{a}{r}.$$

By differentiating, we have
$$\frac{d\theta}{dr} = -\frac{a}{r^2};$$
$$\therefore \frac{r^2 d\theta}{dr} = -a,$$

a constant quantity whatever be the value of r, and consequently when r is infinite. Hence, if PA be the line

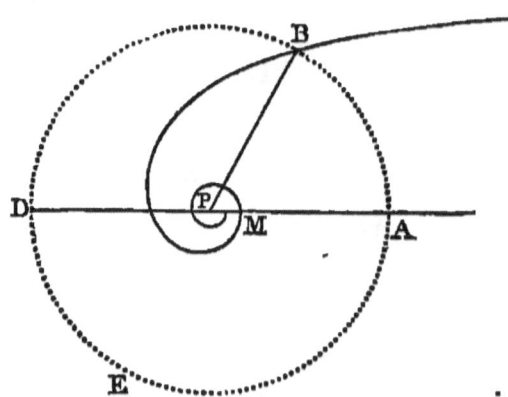

from which θ is measured, it is evidently the direction of the radius vector when infinite; and if a line, PT, were drawn perpendicular to PA and equal to a, then a line,

ASYMPTOTES. 199

TS, drawn through the point T parallel to PA would be the asymptote required.

Ex. 212. Let $r = \dfrac{a}{\sqrt{\theta}}$;

then $r^2 = \dfrac{a^2}{\theta}$,

and $\theta = \dfrac{a^2}{r^2}$.

If $r = \infty$, we shall have $\theta = 0$, and the subtangent
$$\frac{r^2 d\theta}{dr} = -\frac{2a^2}{r},$$
which is also equal to zero when $r = \infty$; hence the line PX, from which the angle θ is estimated, is the asymptote of the spiral.

Ex. 213. Let us take the *Cissoid of Diocles*. Its polar equation is
$$r = \frac{2a \sin.^2 \theta}{\cos. \theta}, \quad (1)$$
where $2a$ is the diameter of the generating circle.

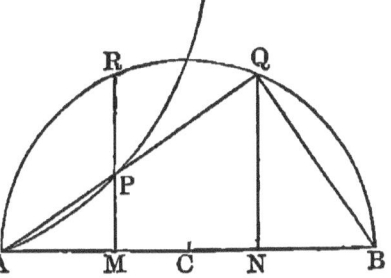

Differentiate; then
$$\frac{dr}{d\theta} = \frac{4a \sin. \theta \cos.^2 \theta + 2a \sin.^3 \theta}{\cos.^2 \theta},$$

or $\dfrac{d\theta}{dr} = \dfrac{\cos.^2 \theta}{4a \sin. \theta \cos.^2 \theta + 2a \sin.^3 \theta}$.

But $r^2 = \dfrac{4a^2 \sin.^4 \theta}{\cos.^2 \theta}$;

$$\therefore r^2 \frac{d\theta}{dr} = \frac{2a \sin.^3 \theta}{2 - \sin.^2 \theta}. \quad (2)$$

If $r = \infty$ in equation (1), we shall have
$$\cos. \theta = 0; \therefore \theta = \tfrac{1}{2}\pi;$$
and in equation (2), the subtangent becomes
$$= 2a = AB.$$

Hence a line drawn perpendicular to the diameter of the circle at B will be an asymptote to the Cissoid.

OSCULATORY CIRCLE.—RADIUS OF CURVATURE.

152. The deviation of a curve from a tangent line is called the *curvature* of that curve; and of several curves, the one which deviates the most rapidly from its tangent is said to have the greatest curvature.

Let AC be a tangent to the curve DBE at the point B, and BM a normal at the same point; then will AC be a tangent to every circle passing through B, and having its centre in the normal BM.

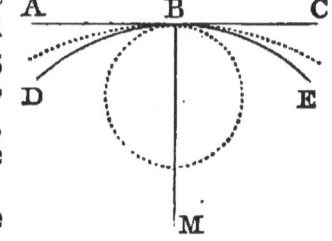

It is evident that the centre of a circle may be taken so near to B that its circumference shall lie within the curve DBE; and it is also evident that the centre may be taken so remote from B that its circumference shall lie between the curve DBE and the tangent AC. Then every circumference described with a greater radius will lie without the curve DBE. Hence there are two classes of tangent circles, the one lying within the curve, the other lying without it; the former having a greater curvature than the curve, the latter a less curvature. Of all the circles that may be described from a centre in the line BM, there is one which coincides most intimately with the curve DBE, and is therefore called the *osculatory circle*, or *circle of curvature*, and its radius is called *the radius of curvature* of the curve.

153. *The curvature of different circumferences is measured by the angle formed by two radii drawn through the extremities of an arc of given length, and this curvature varies inversely as their radii.*

Let r and r' denote the radii of two circles, a the length of a given arc measured on the circumference of each, θ the angle formed by the two radii drawn through the extremities of the arc in the first circle, and θ' the angle formed by the corresponding radii of the second. Then, since the angles are proportional to the arcs which measure them, we shall have the following:

$$\theta : 90° :: a : \tfrac{1}{2}\pi r; \quad \therefore \theta = \frac{180°}{\pi} \cdot \frac{a}{r};$$

OSCULATORY CIRCLE. 201

$$\theta' : 90° :: a : \tfrac{1}{2}\pi r'; \therefore \theta' = \frac{180°}{\pi} \cdot \frac{a}{r'}.$$

Whence $\quad \theta : \theta' :: \dfrac{1}{r} : \dfrac{1}{r'}.$

154. We shall now establish the analytical conditions which make any two curves tangent to each other.

Let PP″ and PP″ be two curves, intersecting each other at P, P″, the co-ordinates of the first curve being x and y, and the co-ordinates of the second being x' and y'.

Then for the common point P we shall have
$x = x'$ and $y = y'$.
Put $\quad\quad RR' = h;$
then, by Taylor's Theorem,

$$P''R' = y + \frac{dy}{dx}h + \frac{d^2y}{dx^2}\frac{h^2}{1.2} +, \text{etc.}; \quad (A)$$

also $\quad P''R' = y' + \dfrac{dy'}{dx'}h + \dfrac{d^2y'}{dx'^2}\dfrac{h^2}{1.2} +,$ etc.; $\quad$ (B)

$$\therefore P''R' - PR = \frac{dy}{dx}h + \frac{d^2y}{dx^2}\frac{h^2}{1.2} +, \text{etc.,}$$

$$P''R' - PR = \frac{dy'}{dx'}h + \frac{d^2y'}{dx'^2}\frac{h^2}{1.2} +, \text{etc.}$$

Equating these values of (P″R′−PR), we shall have, after dividing by h,

$$\frac{dy}{dx} + \frac{d^2y}{dx^2} \cdot \frac{h}{1.2} +, \text{etc.} = \frac{dy'}{dx'} + \frac{d^2y'}{dx'^2}\frac{h}{1.2} +, \text{etc.};$$

and passing to the limit by making $h = 0$,

$$\frac{dy}{dx} = \frac{dy'}{dx'},$$

in which case the point P″ will coalesce with the point P, and the curves will have a common tangent at P.

Two curves which have a common point, and the first differential coefficients of the ordinate at that point equal to each other, are said to have a contact of the *first order*, or are simply tangent to each other.

If the first and second differential coefficients of the

I 2

ordinate at this point are equal to each other respectively, the contact is of the *second order;* and generally,

If the first m differential coefficients of the ordinate at the point of tangency are equal in each curve, the contact is of the *mth order.*

If all the terms in the development (A) were equal to all the terms in (B), the curves would evidently be identical; hence, the greater the number of terms which are equal in the two developments, the more intimate will be the contact of the two curves. Now the circle, from the circumstance of its admitting of all degrees of magnitude by the variation of its radius, from the facility of its description, and the simplicity of its form and properties, has been preferred, to all other curves in which the same degree of osculation may take place, as the osculatory curve. The equation of the circle, being of the second degree, can have only two differential coefficients; hence *the circle will be osculatory to the curve when its first and second differential coefficients are equal to the first and second differential coefficients of the equation of the curve.*

155. The curvature of a curve is thus measured by means of the osculatory circle. Let P, P′ be any two points in the curve, and find the radii, r and r', of the circles which are osculatory at these points; then the

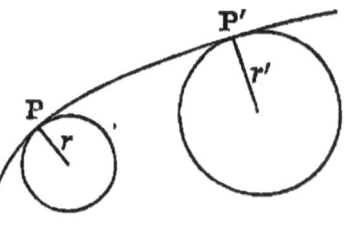

$$\text{curvature at } P : \text{curvature at } P' :: \frac{1}{r} : \frac{1}{r'}.$$

That is, *the curvature at different points of a curve varies inversely as the radius of the osculatory circle.*

156. *To determine the radius of curvature of any proposed curve.*

Let
$$y = f(x); \qquad (1)$$
denote the equation of the proposed curve, and
$$(x-a)^2 + (y-\beta)^2 = R^2, \qquad (2)$$
the equation of the circle.

Then, by differentiation, we obtain
$$(x-a)dx + (y-\beta)dy = 0, \qquad (3)$$
$$dx^2 + dy^2 + (y-\beta)d^2y = 0. \qquad (4)$$

THE RADIUS OF CURVATURE. 203

Combining (2), (3), and (4),
$$y-\beta = -\frac{dx^2+dy^2}{d^2y}, \quad (5)$$

$$x-\alpha = \frac{dy}{dx}\left(\frac{dx^2+dy^2}{d^2y}\right), \quad (6)$$

$$R = \pm\frac{(dx^2+dy^2)^{\frac{3}{2}}}{dx\,d^2y}. \quad (7)$$

From these the radius and co-ordinates of the centre of curvature may be expressed in terms of the co-ordinates of the proposed curve.

In order that the radius of curvature should always be positive, we have only to take the positive sign when the second differential coefficient is positive, and negative when the second differential coefficient is negative.

Ex. 214. Let us take the common parabola whose equation is
$$y^2 = 2px;$$
$$\therefore \frac{dy}{dx} = \frac{p}{y}, \text{ and } \frac{dy^2}{dx^2} = \frac{p^2}{y^2};$$
$$\therefore \frac{d^2y}{dx^2} = -\frac{p^2}{y^3}.$$

Substituting these in (7), taking the negative sign, since the second differential coefficient is negative, we have

$$R = \frac{\left[dx^2\left(1+\frac{p^2}{y^2}\right)\right]^{\frac{3}{2}}}{\frac{p^2 dx^3}{y^3}},$$

which readily reduces to

$$R = \frac{(y^2+p^2)^{\frac{3}{2}}}{p^2},$$

the radius of curvature at any point of the parabola.

(1.) If we make $y=0$, we shall have the radius of curvature at *the vertex* of the parabola.
$$\therefore R = p = \textit{half the parameter}.$$

(2.) If we make $y=p$, we shall have
$$R = 2p\sqrt{2},$$
which is the radius of curvature at the extremity of the *latus rectum*.

157. *To determine the radius of curvature at any point of an ellipse.*

Ex. 215. The equation of the ellipse referred to its centre and axes is
$$a^2y^2 + b^2x^2 = a^2b^2.$$
Differentiating, we obtain
$$\frac{dy}{dx} = -\frac{b^2x}{a^2y}; \quad \therefore \frac{dy^2}{dx^2} = \frac{b^4x^2}{a^4y^2}.$$
$$\frac{d^2y}{dx^2} = -\frac{b^4}{a^2y^3};$$
$$\therefore (dx^2 + dy^2)^{\frac{3}{2}} = \frac{(a^4y^2 + b^4x^2)^{\frac{3}{2}}dx^3}{a^6y^3},$$
and
$$dx\,d^2y = -\frac{b^4dx^3}{a^2y^3}.$$

As the second differential coefficient of y is negative, we take the negative value of R, which makes the radius of curvature positive.

$$\therefore R = -\frac{(dx^2 + dy^2)^{\frac{3}{2}}}{dx\,d^2y},$$
$$= -\frac{(a^4y^2 + b^4x^2)^{\frac{3}{2}}dx^3}{a^6y^3} \div -\frac{b^4dx^3}{a^2y^3},$$
$$= \frac{(a^4y^2 + b^4x^2)^{\frac{3}{2}}}{a^4b^4},$$

which is a general expression for the radius of curvature at any point of the ellipse.

If we take it at the extremity of the transverse axis,
$$x = a \text{ and } y = 0; \quad \therefore R = \frac{b^2}{a}.$$

If we take it at the extremity of the conjugate axis,
$$x = 0 \text{ and } y = b; \text{ then } R = \frac{a^2}{b}.$$

158. To determine the radius of curvature for any proposed curve, we have the following

RULE.

Differentiate the equation of the curve twice; find the values of dy and d^2y, and substitute them in the expression

RADIUS OF CURVATURE.

$$R = \pm \frac{(dx^2 + dy^2)^{\frac{3}{2}}}{dx\,d^2y}.$$

If the radius of curvature for a particular point of the curve be required, then substitute for x and y the coordinates of that point.

159. We will now find the radius of curvature for lines of the second order.
The general equation of these lines is
$$y^2 = mx + nx^2. \tag{1}$$
Differentiating, we have
$$dy = \frac{(m + 2nx)dx}{2y}; \tag{2}$$
$$\therefore dy^2 = \frac{(m + 2nx)^2 dx^2}{4y^2}, \tag{3}$$
and
$$dx^2 + dy^2 = dx^2 + \frac{(m + 2nx)^2 dx^2}{4y^2},$$
$$= \frac{\{4y^2 + (m + 2nx)^2\} dx^2}{4y^2}. \tag{4}$$

Differentiating (2) the second time,
$$d^2y = \frac{2ny\,dx^2 - (m + 2nx)dx\,dy}{2y^2},$$
$$= \frac{\{4ny^2 - (m + 2nx)^2\} dx^2}{4y^3} = -\frac{m^2 dx^2}{4y^3}. \tag{5}$$

Substituting these values in the equation
$$R = -\frac{(dx^2 + dy^2)^{\frac{3}{2}}}{dx\,d^2y},$$
we have
$$R = \frac{\{4(mx + nx^2) + (m + 2nx)^2\}^{\frac{3}{2}}}{2m^2}, \tag{6}$$

which is a general expression for the radius of curvature in lines of the second order for any abscissa x.

If we take it at the extremity of the transverse axis, $x = 0$, and we have
$$R = \tfrac{1}{2}m;$$
that is, in lines of the second order, *the radius of curvature at the vertex of the transverse axis is equal to half the parameter of that axis.*

If it be required to find the radius of curvature at the vertex of the conjugate axis of an ellipse, make

$$m = \frac{2b^2}{a},$$
$$n = -\frac{b^2}{a},$$
$$x = a.$$
Then, by a little reduction, we shall have
$$R = \frac{a^2}{b},$$
as in the preceding problem.

In case of the parabola, in which $n=0$, we have
$$R = \frac{(m^2 + 4mx)^{\frac{3}{2}}}{2m^2},$$
and if $x=0$, $\qquad R = \tfrac{1}{2}m,$
which is the radius of curvature at the vertex of the parabola.

160. *To determine the radius of curvature at any point of the cycloid.*

From the equation
$$dx = \frac{y\,dy}{\sqrt{2ry - y^2}},$$
we have $\qquad dx^2 = \frac{y^2 dy^2}{2ry - y^2}.$

Substitute this value of dx^2 and d^2y in the expression
$$R = -\frac{(dx^2 + dy^2)^{\frac{3}{2}}}{dx\,d^2y},$$
and reduce, we shall obtain
$$R = 2\sqrt{2ry}.$$

Hence the radius of curvature is double the normal (Art. 131).

161. *To determine the radius of curvature at any point of a spiral.*

Let PHD be a spiral, P the pole, DB a tangent, and DG a normal at the point of tangency D. Let C be the centre, and CD the radius of curvature corresponding to the point D. Draw PB and

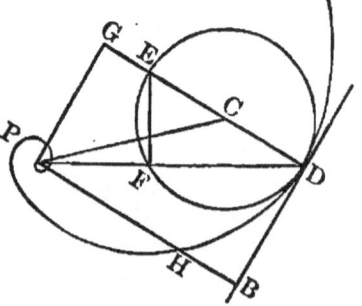

PG respectively perpendicular to DB and DG. Then, considering the point D as describing the circle, the points C and P being fixed, we shall have CD and CP constant.

Put $PD = r =$ radius vector.
$DG = PB = p$,
$CD = \rho =$ radius of curvature.

Then (Geom., B. I., Theorem XXXVII.),

$$\overline{PC}^2 = \overline{PD}^2 + \overline{CD}^2 - 2CD \cdot DG, \qquad (1)$$

or $\qquad \overline{PC}^2 = r^2 + \rho^2 - 2\rho p. \qquad (2)$

Differentiating, we obtain

$$0 = r\,dr - \rho\,dp; \qquad (3)$$

$$\therefore \rho = r\frac{dr}{dp} = \textit{radius of curvature}. \qquad (4)$$

Also from (2) and (4) we have

$$\overline{PC}^2 = r^2 + r^2\frac{dr^2}{dp^2} - 2pr\frac{dr}{dp};$$

$$\therefore PC = \sqrt{r^2 + r^2\frac{dr^2}{dp^2} - 2pr\frac{dr}{dp}}. \qquad (5)$$

Cor. Join FE; then the triangles DPG, DEF are similar, and we have

$$DP : DG :: DE : DF,$$

or $\qquad r : p :: 2r\dfrac{dr}{dp} : 2p\dfrac{dr}{dp} = DF$,

which is the *chord of curvature* at the point D.

Ex. 216. Let us take as an example the logarithmic spiral. By Art. 135, its equation is $\log. r = \theta$.

Hence, by differentiating, we obtain

$$\frac{dr}{d\theta} = \frac{r}{m};$$

and by Art. 152, $\qquad p = \dfrac{r^2}{\sqrt{r^2 + \left(\dfrac{dr}{d\theta}\right)^2}};$

$$\therefore p = \frac{mr}{\sqrt{1 + m^2}};$$

hence $\qquad \rho = \dfrac{r}{m}\sqrt{1 + m^2}.$

And the chord of curvature $= 2p\dfrac{dr}{dp} = 2r$.

Evolutes and Involutes.

162. If a thread fixed around a plane curve be unwound, and kept perfectly tense, its extremity will describe another curve in that plane. The curve from which the thread is unwound is called the *Evolute* of the curve that is formed by the extremity of the moving thread; and this latter curve is called the *Involute*.

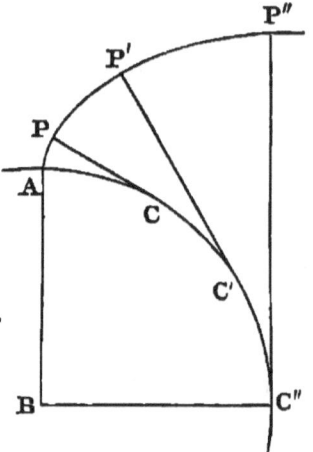

Thus, in the diagram, when the thread kept tense is in the position P″C″, its extremity has described the arc of a curve, APP′P″, the nature of which will depend on the properties of the curve ACC′C″.

From the manner in which the involute is generated from the evolute, we may conclude that the portion of the thread which is disengaged from the evolute *is a tangent to it;* and if a tangent line be drawn to the involute at P, P′, or P″, the line PC, P′C′, or P″C″ would be perpendicular to that tangent at the point of tangency; hence either of these lines may be considered a radius, and the point C, or C′, etc., the centre of a circle of curvature at the corresponding point P, or P′, etc. We may therefore define the evolute of a curve as the *locus of the centres of the circles of curvature at the different points of that curve.*

It is farther evident that *the difference of two radii of curvature is equal to the arc of the evolute intercepted between them,* and that *a tangent to the evolute is a normal to the involute.*

163. *To determine the equation of the evolute of a curve.*

We have seen, Art. 156, that the circle of curvature is characterized by the equations

$$y-\beta = -\frac{dx^2+dy^2}{d^2y}, \qquad (1)$$

EVOLUTE OF A CURVE.

$$x - a = \frac{dy}{dx}\left(\frac{dx^2 + dy^2}{d^2y}\right),$$

or $\qquad x - a = -\frac{dy}{dx}(y - \beta).$ \hfill (2)

If we combine these with the equation of the involute curve, we shall obtain an equation from which x and y may be eliminated.

164. As an illustration, let us take the common parabola as the involute, and thence deduce the equation of its evolute.

Ex. 217. The equation of the parabola is
$$y^2 = 2px. \qquad (1)$$
Differentiating twice, we shall have
$$\frac{dy}{dx} = \frac{p}{y},$$
and $\qquad \dfrac{d^2y}{dx^2} = -\dfrac{p^2}{y^3};$

$$\therefore y - \beta = \frac{(y^2 + p^2)y}{p^2}.$$

Whence $\qquad y^2 = (p^2\beta)^{\frac{2}{3}},$

and $\qquad x - a = -\dfrac{y^2 + p^2}{p}.$

If we substitute $2px$ for y^2, we shall have
$$x - a = -2x - p,$$
or $\qquad x = \frac{1}{3}(a - p).$

Substituting these in (1), we obtain
$$(p^2\beta)^{\frac{2}{3}} = \tfrac{2}{3}p(a - p);$$
$$\therefore \beta^2 = \frac{8}{27p}(a - p)^3,$$

the equation of the evolute.

If we make $\beta = 0$, then $a = p$; hence the evolute meets the axis of x at a distance from the origin equal to half the parameter.

If we remove the origin from A to C, the equation becomes
$$\beta^2 = \frac{8}{27p} a^3,$$
which is called the *semi-cubical parabola*. It passes through the origin C because, when $a = 0$, $\beta = 0$.

The curve consists of two branches symmetrically situated with respect to the axis of x, and lies entirely to the right of the origin, because every positive value of a gives two equal values of β with contrary signs, and every negative value of a makes β impossible.

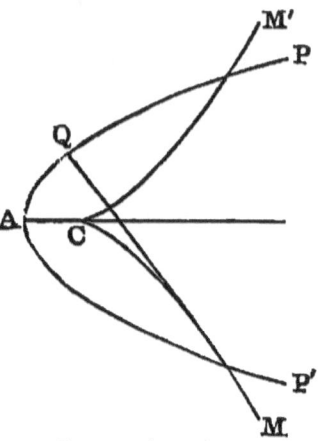

The branch ACM corresponds to the part AP of the involute, and the branch ACM' to the part AP'. The straight line MQ has the same position as an inextensible thread wrapped along ACM, and disengaged from it as in the diagram, at the same time that it is kept constantly tense.

165. Hence, to determine the evolute of a curve, we have the following

RULE.

From the equation of the involute, or given curve, find the values of
$$\frac{dy}{dx} \text{ and } d^2y,$$
and substitute them in the equations
$$y - \beta = -\frac{dx^2 + dy^2}{d^2y},$$
$$x - a = -\frac{dy}{dx}(y - \beta).$$

Combine these equations with the given equation, and eliminate x and y; the resulting equation will be the equation of the evolute.

The only difficulty the student will have to encounter is the elimination of x and y. We may here remark that the number of cases in which this elimination can be readily effected is very small in comparison with the number of curves presented to our notice.

Ex. 218. *To determine the evolute of the ellipse.*

The equation of the ellipse referred to its centre and axes is
$$a^2y^2 + b^2x^2 = a^2b^2.$$

EVOLUTE OF AN ELLIPSE.

Differentiating, we have
$$\frac{dy}{dx} = -\frac{b^2 x}{a^2 y};$$
$$\therefore \frac{dy^2}{dx^2} = \frac{b^4 x^2}{a^4 y^2},$$

and
$$\frac{d^2 y}{dx^2} = -\frac{b^4}{a^2 y^3}.$$

Then
$$y - \beta = -\frac{dx^2 + dy^2}{d^2 y},$$
$$= \frac{(a^4 y^2 + b^4 x^2)y}{a^2 b^4};$$
$$\therefore \beta = y - \frac{(a^4 y^2 + b^4 x^2)y}{a^2 b^4}, \tag{1}$$

and
$$x - a = -\frac{dy}{dx}(y - \beta),$$
$$= \frac{(a^4 y^2 + b^4 x^2)x}{a^4 b^2};$$
$$\therefore a = x - \frac{(a^4 y^2 + b^4 x^2)x}{a^4 b^2}. \tag{2}$$

From the equation of the ellipse, we have
$$a^2 y^2 = a^2 b^2 - b^2 x^2, \text{ or } b^2 x^2 = a^2 b^2 - a^2 y^2.$$
If we multiply the first of these equations by a^2 and the second by b^2, we shall have
$$a^4 y^2 = a^4 b^2 - a^2 b^2 x^2, \text{ and } b^4 x^2 = a^2 b^4 - a^2 b^2 y^2.$$
Adding these two equations member by member, there results
$$a^4 y^2 + b^4 x^2 = a^4 b^2 + a^2 b^4 - a^2 b^2 x^2 - a^2 b^2 y^2, \tag{3}$$
$$= b^2 (a^4 + a^2 b^2 - a^2 x^2 - a^2 y^2). \tag{4}$$
Putting for $a^2 b^2$ its value $a^2 y^2 + b^2 x^2$, we have
$$a^4 y^2 + b^4 x^2 = b^2 (a^4 + a^2 y^2 + b^2 x^2 - a^2 x^2 - a^2 y^2),$$
$$= b^2 \{a^4 - (a^2 - b^2) x^2\},$$
$$= b^2 (a^4 - c^2 x^2); \because c^2 = a^2 - b^2. \tag{5}$$
Again: taking (Eq. 3), we shall have
$$a^4 y^2 + b^4 x^2 = a^2 (a^2 b^2 + b^4 - b^2 x^2 - b^2 y^2),$$
$$= a^2 (a^2 y^2 + b^2 x^2 + b^4 - b^2 x^2 - b^2 y^2),$$
$$= a^2 \{b^4 + (a^2 - b^2) y^2\},$$
$$= a^2 (b^4 + c^2 y^2). \tag{6}$$
Substituting these values of $a^4 y^2 + b^4 x^2$ in (Eqs. 1 and 2), we have

$$\beta = y - \frac{a^2 y (b^4 + c^2 y^2)}{a^2 b^4}, \quad (7)$$

$$a = x - \frac{b^2 x (a^4 - c^2 x^2)}{a^4 b^2}. \quad (8)$$

Finding the values of y and x from (7) and (8) in terms of a and β, and substituting them in the equation of the ellipse, we obtain

$(b\beta)^{\frac{2}{3}} + (aa)^{\frac{2}{3}} = c^{\frac{4}{3}}$, *the equation of the evolute.*

If $a = 0$, then $\beta = \pm \frac{c^2}{b}$; hence the curve meets the axis of y in two points equidistant from the origin.

If $\beta = 0$, then $a = \pm \frac{c^2}{a}$; hence the curve meets the axis of x in two points equidistant from the origin also.

Ex. 219. To determine the radius vector in the spiral of Archimedes when the radius of curvature is equal to the chord of curvature.

The equation to the curve is (Art. 134)
$$r = a\theta;$$
$$\therefore \frac{dr}{d\theta} = a.$$

But $\quad \dfrac{dr}{d\theta} = \dfrac{r\sqrt{r^2 - p^2}}{p}$ (Art. 149);

$$\therefore \frac{r\sqrt{r^2 - p^2}}{p} = a, \text{ or } r^4 - p^2 r^2 = a^2 p^2;$$

$$\therefore 4r^3 \frac{dr}{dp} - 2pr^2 - 2p^2 r \frac{dr}{dp} = 2a^2 p,$$

whence $\quad r\dfrac{dr}{dp} = \dfrac{p(r^2 + a^2)}{2(2r^2 - p^2)}.$

But $\quad \dfrac{r^4}{p^2} = r^2 + a^2;$

$$\therefore p^2 = \frac{r^4}{r^2 + a^2}, \text{ or } p = \frac{r^2}{(r^2 + a^2)^{\frac{1}{2}}}.$$

Hence $\quad \rho = r\dfrac{dr}{dp} = \dfrac{(r^2 + a^2)^{\frac{3}{2}}}{2(r^2 + 2a^2)} =$ rad. of curvature.

Now the chord of curvature $= 2p\dfrac{dr}{dp} = \dfrac{2r(r^2 + a^2)}{2(r^2 + 2a^2)}.$

EVOLUTE OF THE RECTANGULAR HYPERBOLA. 213

Comparing this value of the chord of curvature with the value of the radius of curvature, it appears that the radius of curvature is equal to the chord if

$$(r^2+a^2)^{\frac{3}{2}}=2r(r^2+a^2), \text{ or } (r^2+a^2)^{\frac{1}{2}}=2r,$$

from which we determine the *radius vector*

$$r=\frac{a}{\sqrt{3}}.$$

Ex. 220. To determine the evolute of the rectangular hyperbola referred to its asymptotes from the equation
$$xy=a^2.$$
By differentiation we obtain
$$\frac{dy}{dx}=-\frac{y}{x}=-\frac{a^2}{x^2};$$
$$\therefore \frac{dy^2}{dx^2}=\frac{a^4}{x^4}.$$
Also
$$\frac{d^2y}{dx^2}=\frac{2a^2}{x^3};$$
$$\therefore y-\beta=-\frac{x^3}{2a^2}-\frac{a^2}{2x}, \qquad (1)$$
$$x-a=-\frac{x}{2}-\frac{a^4}{2x^3}, \qquad (2)$$
or
$$\beta=\frac{3a^2}{2x}+\frac{x^3}{2a^2}, \qquad (3)$$
$$a=\frac{3x}{2}+\frac{a^4}{2x^3}. \qquad (4)$$

By addition and subtraction we have
$$a+\beta=\frac{a^4}{2x^3}+\frac{3a^2}{2x}+\frac{3x}{2}+\frac{x^3}{2a^2},$$
$$=\frac{a}{2}\left(\frac{a^3}{x^3}+3\frac{a}{x}+3\frac{x}{a}+\frac{x^3}{a^3}\right),$$
$$=\frac{a}{2}\left(\frac{a}{x}+\frac{x}{a}\right)^3; \qquad (5)$$
$$a-\beta=\frac{a}{2}\left(\frac{a^3}{x^3}-3\frac{a}{x}+3\frac{x}{a}-\frac{x^3}{a^3}\right),$$
$$=\frac{a}{2}\left(\frac{a}{x}-\frac{x}{a}\right)^3; \qquad (6)$$
$$\therefore (a+\beta)^{\frac{1}{3}}=\left(\frac{a}{2}\right)^{\frac{1}{3}}\left(\frac{a}{x}+\frac{x}{a}\right),$$

or
$$(a+\beta)^{\frac{2}{3}}=\left(\frac{a}{2}\right)^{\frac{2}{3}}\left(\frac{a}{x}+\frac{x}{a}\right)^2; \qquad (7)$$

and
$$(a-\beta)^{\frac{1}{3}}=\left(\frac{a}{2}\right)^{\frac{1}{3}}\left(\frac{a}{x}-\frac{x}{a}\right),$$

or
$$(a-\beta)^{\frac{2}{3}}=\left(\frac{a}{2}\right)^{\frac{2}{3}}\left(\frac{a}{x}-\frac{x}{a}\right)^2. \qquad (8)$$

Whence $(a+\beta)^{\frac{2}{3}}-(a-\beta)^{\frac{2}{3}}=4\left(\frac{a}{2}\right)^{\frac{2}{3}}=(4a)^{\frac{2}{3}}$,

the equation of the evolute.

Ex. 221. To determine the radius of curvature of an ellipse in terms of the angle which the normal makes with the transverse axis.

The normal $PN = y\sqrt{1+\dfrac{dy^2}{dx^2}}$.

Put the angle $PNR = \theta$;

then $\sin \theta = \dfrac{PR}{PN} = \dfrac{y}{PN}$;

$\therefore \sin \theta = \dfrac{1}{\sqrt{1+\dfrac{dy^2}{dx^2}}}$.

But $y = \dfrac{B}{A}\sqrt{A^2 - x^2}$,

the equation of the ellipse.

$$\therefore \frac{dy}{dx} = -\frac{Bx}{A\sqrt{A^2-x^2}},$$

and
$$1+\frac{dy^2}{dx^2} = 1+\frac{B^2x^2}{A^2(A^2-x^2)} = \frac{A^2-e^2x^2}{A^2-x^2};$$

$$\therefore \left(1+\frac{dy^2}{dx^2}\right)^{\frac{3}{2}} = \left(\frac{A^2-e^2x^2}{A^2-x^2}\right)^{\frac{3}{2}},$$

and
$$\frac{d^2y}{dx^2} = -\frac{AB}{(A^2-x^2)^{\frac{3}{2}}}.$$

Hence
$$R = -\frac{\left(1+\dfrac{dy^2}{dx^2}\right)^{\frac{3}{2}}}{\dfrac{d^2y}{dx^2}} = \frac{(A^2-e^2x^2)^{\frac{3}{2}}}{AB} \qquad (A)$$

But $\sin.^2 \theta = \dfrac{1}{1+\dfrac{dy^2}{dx^2}} = \dfrac{A^2-x^2}{A^2-e^2x^2}$;

$\therefore A^2 \sin.^2 \theta - e^2x^2 \sin.^2 \theta = A^2 - x^2$,

or $x^2 = \dfrac{A^2(1-\sin.^2 \theta)}{1-e^2 \sin.^2 \theta}$,

$e^2x^2 = \dfrac{A^2e^2(1-\sin.^2 \theta)}{1-e^2 \sin.^2 \theta}$,

$A^2 - e^2x^2 = A^2 - \dfrac{A^2e^2(1-\sin.^2 \theta)}{1-e^2 \sin.^2 \theta} = \dfrac{A^2(1-e^2)}{1-e^2 \sin.^2 \theta}$.

Substituting this in Eq. A and reducing, we have

$$R = \dfrac{A(1-e^2)}{(1-e^2 \sin.^2 \theta)^{\frac{3}{2}}}.$$

If $\theta = 0$ or $180°$, $R = \dfrac{B^2}{A}$; if $\theta = 90°$, $R = \dfrac{A^2}{B}$, as in Ex. 215.

CHAPTER IV.

ON VANISHING FRACTIONS.—MAXIMA AND MINIMA OF FUNCTIONS OF ONE AND TWO VARIABLES.—CHANGING THE INDEPENDENT VARIABLE.—TANGENT PLANES AND NORMAL LINES TO CURVED SURFACES.—RADIUS OF CURVATURE OF CURVES OF DOUBLE CURVATURE.

166. It was shown in Algebra (Art. 87) that the expression $\dfrac{0}{0}$ is not always a symbol of indetermination, but that it sometimes denotes the presence of a common factor, which being canceled, the true value of the fraction may be obtained. We shall now proceed to find the value of any fraction whose numerator and denominator both vanish for some particular value of the variable.

Let $\dfrac{P(x-a)^m}{Q(x-a)^n}$

be the form of the fraction, in which P and Q are functions of x, which do not reduce to zero when $x = a$. We shall have, if we substitute at first,

$$\frac{P(x-a)^m}{Q(x-a)^n} = \frac{0}{0}.$$

The value of the fraction, however, may be *zero, finite* or *infinite*, depending upon the values of m and n.

Thus, if m be greater than n,
$$\frac{P(x-a)^{m-n}}{Q} = \frac{0}{Q} = 0.$$

If m be less than n,
$$\frac{P}{Q(x-a)^{n-m}} = \frac{P}{0} = \text{infinity}.$$

If m be equal to n, then
$$\frac{P}{Q} = \text{a finite quantity}.$$

Put X equal to the numerator, and X' equal to the denominator of the proposed fraction, each being such a function of x that for some particular value of x they both reduce to zero; then give to x an increment, h, and develop by *Taylor's Theorem*, and we have

$$\frac{X + \dfrac{dX}{dx}h + \dfrac{d^2X}{dx^2}\dfrac{h^2}{1.2} + \dfrac{d^3X}{dx^3}\dfrac{h^3}{1.2.3} +, \text{etc.}}{X' + \dfrac{dX'}{dx}h + \dfrac{d^2X'}{dx^2}\dfrac{h^2}{1.2} + \dfrac{d^3X'}{dx^3}\dfrac{h^3}{1.2.3} +, \text{etc.}}$$

If we substitute for x its value a, both X and X' reduce to zero; then, dividing both numerator and denominator by h, we have

$$\frac{\dfrac{dX}{dx} + \dfrac{d^2X}{dx^2}\dfrac{h}{1.2} + \dfrac{d^3X}{dx^3}\dfrac{h^2}{1.2.3} +, \text{etc.}}{\dfrac{dX'}{dx} + \dfrac{d^2X'}{dx^2}\dfrac{h}{1.2} + \dfrac{d^3x}{dx^3}\dfrac{h^2}{1.2.3} +, \text{etc.}}$$

Now, making $h=0$, this fraction reduces to $\dfrac{\frac{dX}{dx}}{\frac{dX'}{dx}}$; and if, by substituting for x its value a, we find that each term of this fraction becomes 0, we must expunge these terms, and divide again by h both numerator and denominator, until we obtain a fraction of which the numerator and denominator do not both reduce to zero. Hence we obtain the following

RULE.

Find the differential coefficients of the numerator and denominator separately, and if one or the other of these do not reduce to zero when the value of x is substituted, the ratio of these differential coefficients will be the value of the fraction; but if both reduce to zero, we must differentiate again, until we find a differential coefficient which does not reduce to zero when the value of x is substituted for x; then the ratio of these differential coefficients will be the value of the vanishing fraction.

Examples.

Ex. 222. Find the value of the fraction
$$\frac{\tan. x - x}{x - \sin. x} \text{ when } x = 0.$$
If we substitute for x its value $=0$, we shall have
$$\frac{\tan. 0 - 0}{0 - \sin. 0} = \frac{0}{0}.$$

Put $\qquad u = \tan. x - x;$

then $\qquad \dfrac{du}{dx} = (1 + \tan.^2 x) - 1;$

and when $x=0,\ \dfrac{du}{dx} = 0,$

and $\qquad u' = x - \sin. x;$

then $\qquad \dfrac{du'}{dx} = 1 - \cos. x;$

and when $x=0,\ \dfrac{du'}{dx} = 0.$

Differentiating the second time, we have
$$\dfrac{d^2u}{dx^2} = 2 \tan. x(1 + \tan.^2 x);$$

and when $x=0,\ \dfrac{d^2u}{dx^2} = 0;$

$$\dfrac{d^2u'}{dx^2} = \sin. x;$$

and when $x=0,\ \dfrac{d^2u'}{dx^2} = 0.$

Differentiating the third time, we have

K

$$\frac{d^3u}{dx^3} = 2(1+\tan.^2 x)^2 + 4\tan.^2 x\,(1+\tan.^2 x);$$

and when $x=0$, $\quad \dfrac{d^3u}{dx^3}=2,$

$$\frac{d^3u'}{dx^3}=\cos. x\,;$$

and when $x=0$, $\quad \dfrac{d^3u'}{dx^3}=1\,;$

$$\therefore \frac{\dfrac{d^3u}{dx^3}}{\dfrac{d^3u'}{dx^3}} = \frac{2}{1} = 2 = \text{value of the fraction.}$$

Ex. 223. Find the value of the fraction
$$\frac{\log. x}{x-1} \text{ when } x=1.$$

Put $\quad u=\log. x\,;\ \therefore \dfrac{du}{dx}=\dfrac{1}{x}\,;$

and when $x=1$, $\quad \dfrac{du}{dx}=1,$

$u'=x-1\,;\ \therefore \dfrac{du'}{dx}=1\,;$

$$\therefore \frac{\dfrac{du}{dx}}{\dfrac{du'}{dx}} = 1 = \text{the value of the fraction.}$$

Ex. 224. Find the value of the fraction
$$\frac{x-1}{x^3-1} \text{ when } x=1.$$

Ans. $\frac{1}{3}$.

Ex. 225. Find the value of the fraction
$$\frac{x-1}{x^n-1} \text{ when } x=1.$$

Ans. $\dfrac{1}{n}$.

Ex. 226. Find the value of the fraction
$$\frac{a^n-x^n}{\log. a - \log. x} \text{ when } x=a.$$

Ans. na^n.

VANISHING FRACTIONS. 219

Ex. 227. Find the value of the fraction
$$\frac{1-x}{\cot \frac{1}{2}\pi x} \text{ when } x=1.$$

Ans. $\frac{2}{\pi}$.

167. In the last article the numerator and denominator of the fraction were developed by means of *Taylor's Theorem;* but when this development can not be made by that Theorem, recourse must be had to the common algebraic methods of obtaining their expansions. Thus, if
$$u=\frac{f(x)}{f'(x)},$$
where $f(x)$ and $f'(x)$ each become $=0$ by assigning to x the particular value a, let $a+h$ be substituted for x; and suppose that by the operations of Algebra we obtain
$$u'=\frac{Ah^{a}+Bh^{\beta}+, \text{ etc.}}{A'h^{a'}+B'h^{\beta'}+, \text{ etc.}}$$
then, if a be greater than a', we shall have, by dividing both terms by $h^{a'}$,
$$u'=\frac{Ah^{a-a'}+Bh^{\beta-a'}+, \text{ etc.}}{A'+B'h^{\beta'-a'}+, \text{ etc.}}.$$
Now if $h=0$, we shall get
$$u=0.$$
Again: if $a=a'$, we obtain, by dividing by h^{a},
$$u'=\frac{A+Bh^{\beta-a}+, \text{ etc.}}{A'+B'h^{\beta'-a}+, \text{ etc.}},$$
where, if $h=0$, we have
$$u=\frac{A}{A'}=a \text{ finite quantity.}$$

Also, if a be less than a', then, by dividing by h^{a}, we obtain
$$u'=\frac{A+Bh^{\beta-a}+, \text{ etc.}}{A'h^{a'-a}+B'h^{\beta'-a}+, \text{ etc.}},$$
which, on the supposition that $h=0$, gives the corresponding value of
$$u=\infty.$$

Hence the particular value of the fraction will be *zero, finite* or *infinite*, according as the least exponent of the development of the numerator is greater than, equal to, or less than that of the denominator. We may remark that this method is applicable in all cases, and may frequently be used with advantage even in those examples in which Art. 166 applies.

Ex. 229. Find the value of the fraction
$$\frac{x^x-x}{\log.x-x+1} \text{ when } x=1.$$
If we substitute *unity* for x, this fraction becomes
$$\frac{0}{0}.$$
Since the value of x that makes this fraction evanescent is *unity* instead of a, we put $1+h$ for x; then
$$\frac{(1+h)^{1+h}-(1+h)}{\log.(1+h)-h}.$$
Developing the terms of the fraction, we have
$$\frac{h^2+\frac{1}{2}(1+h)h^3+, \text{etc.}}{-\frac{1}{2}h^2+\frac{1}{3}h^3-\frac{1}{4}h^4+, \text{etc.}}.$$
Divide by h^2; then
$$\frac{1+\frac{1}{2}(1+h)h+, \text{etc.}}{-\frac{1}{2}+\frac{1}{3}h-\frac{1}{4}h^2+, \text{etc.}}.$$
Now if h be made equal to 0, the value of the fraction is -2.

This example may also be solved by the first Rule by differentiating twice.

Ex. 230. Find the value of the fraction
$$\frac{(x^2-3ax+2a^2)^{\frac{2}{3}}}{(x^3-a^3)^{\frac{1}{2}}} \text{ when } x=a.$$
Ans. 0.

Ex. 231. Find the value of the fraction
$$\frac{(x^2-a^2)^{\frac{3}{2}}}{(x-a)^{\frac{3}{2}}} \text{ when } x=a.$$
Ans. $(2a)^{\frac{3}{2}}$.

168. The method of determining the value of a fraction whose numerator and denominator become *zero* for some particular value of the variable, leads us to obtain the value of any fraction whose numerator and denominator become *infinite* for some particular value of the variable.

VANISHING FRACTIONS.

Let $f(x)$ and $f'(x)$ denote two different functions of x, such that the fraction
$$\frac{f(x)}{f'(x)}$$
becomes $\frac{\infty}{\infty}$ when $x=a$; then it is obvious that
$$\frac{f(x)}{f'(x)} = \frac{\dfrac{1}{f'(x)}}{\dfrac{1}{f(x)}} = \frac{0}{0}.$$

So that if we determine by either of the preceding methods the value of this last fraction, the value of the proposed fraction will also be obtained.

Ex. 232. Required the value of
$$\frac{\tan.\dfrac{\pi x}{2a}}{a^{-1}(x^2-a^2)^{-1}x^2} \quad \text{when } x=a.$$

This may be put under the form
$$\frac{a(x^2-a^2)x^{-2}}{\cot.\dfrac{\pi x}{2a}}. \qquad Ans. \ -\frac{4a}{\pi}.$$

169. In like manner we may find the true value of the *product* of two functions, one of which becomes *zero*, the other *infinite* for some particular value of the variable.

For, let $f(x)$ and $f'(x)$ be the functions. Then, if by substituting the value a for x in each of these functions, we have
$$f(x)=0 \text{ and } f'(x)=\infty, \text{ or } f(x)\,f'(x)=0\times\infty, \quad (1)$$
their product may be put in the form
$$\frac{f(x)}{\dfrac{1}{f'(x)}} = \frac{0}{0}. \qquad (2)$$

170. Again: the true value of the *difference* of two functions may be obtained in the case in which the substitution of a particular value for the variable causes each of them to become *infinite*.

Thus, if $\quad f(x)=\infty, \text{ and } f'(x)=\infty,$
then $\qquad\qquad f(x)-f'(x)=\infty-\infty \qquad (1)$
may be put in the form

$$\frac{\dfrac{1}{f'(x)} - \dfrac{1}{f(x)}}{\dfrac{1}{f(x)\,f'(x)}} = \frac{0}{0}. \qquad (2)$$

Ex. 231. Find the value of the function

$$a(1-x)\tan.\frac{\pi x}{2} \text{ when } x=1.$$

Here the first factor $a(1-x)$ becomes *zero*, and the second *infinite* when $x=1$; but (Art. 169)

$$a(1-x)\tan.\frac{\pi x}{2} = \frac{a(1-x)}{\cot.\dfrac{\pi x}{2}}.$$

Put $\qquad u = a(1-x);$

then $\qquad \dfrac{du}{dx} = -a.$

Put $\qquad u' = \cot.\tfrac{1}{2}\pi x;$

then $\qquad \dfrac{du'}{dx} = -\dfrac{\pi}{2}\left(1+\cot.^2\dfrac{\pi x}{2}\right);$

$$\therefore \frac{\dfrac{du}{dx}}{\dfrac{du'}{dx}} = \frac{a}{\dfrac{\pi}{2}\left(1+\cot.^2\dfrac{\pi x}{2}\right)};$$

and when $x=1$, $\qquad = \dfrac{2a}{\pi}.$

Ex. 232. Find the value of the function

$$\frac{1}{\log. x} - \frac{x}{\log. x} \text{ when } x=1.$$

This becomes $\infty - \infty$. But, by (Art. 172),

$$\frac{1}{\log. x} - \frac{x}{\log. x} = \frac{1-x}{\log. x} = \frac{0}{0} \text{ when } x=1.$$

Put $\qquad u = 1-x;$ then $\dfrac{du}{dx} = -1.$

$\qquad u' = \log. x,\qquad \dfrac{du'}{dx} = \dfrac{1}{x};$

$$\therefore \frac{\dfrac{du}{dx}}{\dfrac{du'}{dx}} = -1 = \text{value required.}$$

VANISHING FRACTIONS. 223

Miscellaneous Examples.

Ex. 233. Find the value of the fraction
$$\frac{a^{\log. x} - x}{\log. x} \text{ when } x = 1.$$
$$Ans.\ l\left(\frac{a}{e}\right).$$

Ex. 234. Find the value of the fraction
$$\frac{e^x - 1 - l(1+x)}{x^2}$$
when $x = 0$, and e the base of the Naperian log.
$$Ans.\ 1.$$

Ex. 235. Find the value of the fraction
$$\frac{a^n - x^n}{\log. (a)^n - \log. (x)^n} \text{ when } x = a.$$
$$Ans.\ a^n.$$

Ex. 236. Find the value of the fraction
$$\frac{e^{2x} - 1}{e^x \log. (1+x)} \text{ when } x = 0.$$
$$Ans.\ 2.$$

Ex. 237. Find the value of the fraction
$$\frac{2}{x^2 - 1} - \frac{1}{x - 1} \text{ when } x = 1.$$
$$Ans.\ -\tfrac{1}{2}.$$

Ex. 238. Find the value of the fraction
$$\frac{a - (a^2 - x^2)^{\frac{1}{2}}}{x^2} \text{ when } x = 0.$$
$$Ans.\ \frac{1}{2a}.$$

Ex. 239. Find the value of the fraction
$$\frac{(a+x)^{\frac{1}{2}} - (2a)^{\frac{1}{2}}}{(a+2x)^{\frac{1}{2}} - (3a)^{\frac{1}{2}}} \text{ when } x = a.$$
$$Ans.\ \tfrac{1}{4}\sqrt{6}.$$

Ex. 240. Find the value of the fraction
$$\frac{xe^{2x} + 1 - e^{2x} - x}{e^{2x} - 1} \text{ when } x = 0.$$
$$Ans.\ -1.$$

ON THE MAXIMA AND MINIMA OF FUNCTIONS OF ONE VARIABLE.

171. Definition.—*A maximum value of a variable function is that value which is greater than the value which immediately precedes or immediately follows it. And the minimum value of a variable function is less than that which immediately precedes or immediately follows it.*

If $u = f(x)$,
and let x become $x+h$, and $x-h$; then the corresponding values of the function will be

$$u = f(x), \qquad (1)$$
$$u' = f(x+h), \qquad (2)$$
$$u'' = f(x-h). \qquad (3)$$

Then, by Taylor's Theorem, we shall have

$$u' = f(x+h) = u + \frac{du}{dx}h + \frac{d^2u}{dx^2}\frac{h^2}{1.2} + \frac{d^3u}{dx^3}\frac{h^3}{1.2.3} +, \text{etc.} \qquad (4)$$

and

$$u'' = f(x-h) = u - \frac{du}{dx}h + \frac{d^2u}{dx^2}\frac{h^2}{1.2} - \frac{d^3u}{dx^3}\frac{h^3}{1.2.3} +, \text{etc.} \qquad (5)$$

Now, if u is a maximum, we must have from the definition u' and u'' both less than u; or, if u is a minimum, we must have u' and u'' both greater than u, however small h may be assumed. By (Art. 56), h can be taken of such magnitude that $\frac{du}{dx}h$ shall be greater than the sum of all the terms which follow it. Now, in equation (4), u is less than u', and in (5) u is greater than u'', which can not be the case if u is a maximum or a minimum; and therefore, when u is either a maximum or a minimum,

$$\frac{du}{dx} = 0.$$

When the first differential coefficient vanishes,

$$u' = u + \frac{d^2u}{dx^2}\frac{h^2}{1.2} + \frac{d^3u}{dx^3}\frac{h^3}{1.2.3} +, \text{etc.,}$$

and

$$u'' = u + \frac{d^2u}{dx^2}\frac{h^2}{1.2} - \frac{d^3u}{dx^3}\frac{h^3}{1.2.3} +, \text{etc.,}$$

which, on the principle that h can be taken so small that

$\dfrac{d^2u}{dx^2} \dfrac{h^2}{1.2}$ shall be greater than the sum of all the succeeding terms, u' and u'' are both greater than u when $\dfrac{d^2u}{dx^2}$ is positive, and both less than u when $\dfrac{d^2u}{dx^2}$ is negative. That is, u will be a maximum when the values of x, obtained from the equation $\dfrac{du}{dx}=0$, renders the second differential coefficient *negative*, and a minimum when these values make the second differential coefficient *positive*.

It may happen that the roots of the equation
$$\dfrac{du}{dx}=0,$$
when substituted in the second differential coefficient, cause that coefficient to vanish. The signs of the developments will then depend on the signs of the third differential coefficients; and as these signs are different, we must have
$$\dfrac{d^3u}{dx^3}=0.$$
The conditions of maximum and minimum will then be, that if the values of x obtained from the equation
$$\dfrac{du}{dx}=0,$$
being substituted in the *fourth* differential coefficient, make that coefficient negative, it will be a *maximum*, and if they make it positive, a *minimum*. Hence we derive the following

RULE.

Differentiate the function; put the first differential coefficient equal to zero, and find the roots of that equation. Then substitute these roots in succession in the succeeding differential coefficients, stopping at the first that does not vanish. If this is of an odd order, the values that we have used will not render the function either a maximum or a minimum; but if it is of an even order and negative, the function will be a maximum, if positive, a minimum.

K 2

172. Before we apply the preceding theory to examples, it is proper to state that if the given function has a *constant* factor, that factor may be omitted.

If a function is a maximum or a minimum, its square, cube, or any other power must obviously be a maximum or minimum also. Hence, when the function is under a radical sign, that sign may be removed. The rational function may, however, become a maximum or a minimum for more values of the variable than the original root; indeed, all values of the variable which render the rational function *negative*, will render every even root of it imaginary; such values, therefore, do not belong to that root.

Examples.

Ex. 241. Let $u = x(a-x)^2$; to find the value of x which will render u a maximum or a minimum.

Differentiating, we have
$$\frac{du}{dx} = (a-x)^2 - 2x(a-x),$$
and
$$\frac{d^2u}{dx^2} = 6x - 4a.$$

Now, if u be a maximum or a minimum, we must have
$$\frac{du}{dx} = 0,$$
or $\qquad (a-x)^2 - 2x(a-x) = 0;$
$$\therefore x = \tfrac{1}{3}a, \text{ or } x = a.$$

Substitute $\tfrac{1}{3}a$ for x in the second differential coefficient; then
$$\frac{d^2u}{dx^2} = -2a.$$
And substituting a for x in the same, we have
$$\frac{d^2u}{dx^2} = 2a.$$

The value $x = \tfrac{1}{3}a$ renders u a maximum, and $x = a$ makes it a minimum.

Ex. 242. Find the value of θ which will render the function $u = \sin.\theta - \text{vers}.\theta$ a maximum.
$$\frac{du}{d\theta} = \cos.\theta - \sin.\theta,$$
and
$$\frac{d^2u}{d\theta^2} = -\sin.\theta - \cos.\theta.$$

Putting the first differential coefficient equal to 0,
$$\cos \theta - \sin \theta = 0;$$
$$\therefore \sin \theta = \cos \theta = \sin (90° - \theta);$$
$$\therefore \theta = 90° - \theta; \therefore \theta = 45°.$$

Substituting the value of θ in the second differential coefficient,
$$\frac{d^2 u}{d\theta^2} = -\tfrac{1}{2}\sqrt{2} - \tfrac{1}{2}\sqrt{2},$$
$$= -\sqrt{2}.$$

The second differential coefficient being negative, proves that u is a *maximum*.

Ex. 243. Find the values of x that will render u a maximum or minimum in the equation
$$u = x^4 - 16x^3 + 88x^2 - 192x + 150.$$

Differentiate; then
$$\frac{du}{dx} = 4x^3 - 48x^2 + 176x - 192.$$

Equating this with zero, we have
$$4x^3 - 48x^2 + 176x - 192 = 0,$$
$$x^3 - 12x^2 + 44x - 48 = 0,$$
an equation of the third degree, whose roots are 2, 4, and 6. If we differentiate again, we shall have
$$\frac{d^2 u}{dx^2} = 12x^2 - 96x + 176.$$

Substituting 2 for x in this second differential coefficient, the result will become $+32$, which shows that u is a minimum; substituting 4 for x, we have -16: this being negative, indicates a maximum. Again: substituting 6 for x, we obtain $+32$, a minimum. The adjoining diagram represents the form of the curve. There are two minima values corresponding to the abscissas 2 and 6, and one maximum value corresponding to the abscissa 4.

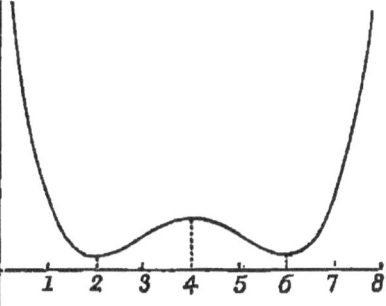

Ex. 244. Find the values of x which render
$$u = (3a^2 x - x^3)^{\frac{1}{3}}$$
a maximum or a minimum.

Ans. $+a$, a *max.*
$-a$, a *min.*

Ex. 245. Find the value of x which will render
$$u = \frac{(x-1)^m}{(x+1)^n}$$
a maximum or a minimum, m and n being positive whole numbers, and $n > m$.

Ans. $x = \dfrac{n+m}{n-m} =$ a *max.*

But if $n < m$, then this value of x will render u a *maximum* or a *minimum*, according as $m-n$ is an *odd* or *even* number.

Ex. 246. Find the values of x which will render
$$u = e^x \cos.(x-a)$$
a maximum or a minimum.

Ans. $x = \frac{1}{4}\pi + a =$ a *max.*
$x = \frac{5}{4}\pi + a =$ a *min.*

Ex. 247. Find the values of x which will render
$$u = \frac{e^x}{\sin.(x-a)}$$
a maximum or a minimum.

Ans. $x = \frac{1}{4}\pi + a =$ a *min.*
$x = \frac{5}{4}\pi + a =$ a *max.*

Ex. 248. Find the value of x which will render
$$u = x^x$$
a maximum or a minimum.

Ans. $x = \dfrac{1}{e} =$ a *min.*

Ex. 249. *Find the maximum rectangle which can be inscribed in a given triangle.*

Ans. Alt. of rec. = half the height of the $\triangle$.

Ex. 250. *Given the base and perpendicular of a plane triangle, to find the other two sides when the vertical angle is a maximum.*   *Ans.* The $\triangle$ is isosceles.

Ex. 251. *Given the hypothenuse of a right angled triangle, to find the other sides when the area is a maximum.*   *Ans.* The two sides are equal.

MAXIMA AND MINIMA OF FUNCTIONS. 229

Ex. 252. *Given the volume of a cylinder* $=a^3$, *to find the altitude and radius of the base when the surface is a minimum.*

$$Ans.\ x = \frac{a}{(2\pi)^{\frac{1}{3}}} = rad.$$

$$y = 2\frac{a}{(2\pi)^{\frac{1}{3}}} = alt.$$

Ex. 253. *Cut the greatest parabola from a given right cone.*

Ans. Alt. par. $=\frac{3}{4}$ slant height of the cone.

Ex. 254. *Inscribe the greatest rectangle in a given ellipse.*

Ans. The sides $=a\sqrt{2}$ and $b\sqrt{2}$, where a and b are the semi-axes.

Ex. 255. Find the value of x which will render
sin. x $(1 + \cos. x)$ a maximum.

Ans. $x = 60°$.

Ex. 256. Find the altitude of the greatest right cone that can be cut out of a given ellipsoid of revolution.

Ans. Alt. $=\frac{4}{3}$ axis.

Ex. 257. Find the altitude of the greatest cylinder that can be cut out of a given paraboloid.

Ans. Alt. $=\frac{1}{2}$ axis.

Ex. 258. Find the least triangle that can be formed by the radii produced, and a tangent to the quadrant of a circle.

Ans. The point of tangency is at the middle of the arc.

Ex. 259. Within a given parabola to inscribe the greatest parabola, the vertex of the latter being at the middle point of the base of the former.

Ans. The axis $=\frac{2}{3}$ of the axis of the given par.

Ex. 260. Inscribe the greatest cylinder within a given right cone.

Ex. 261. Inscribe the greatest rectangle in a given segment of a circle.

Ex. 262. Find the value of x which will render

$$\frac{\sin. x}{1 + \tan. x}\ \text{a maximum.}$$

Ans. $x = 45°$.

Ex. 263. Find the value of x which shall render $\sin.^3 x \cos. x$ a maximum.

$Ans.\ x=60°.$

Ex. 264. Find the least parabola which shall circumscribe a given circle.

$Ans.$ Alt.$=\frac{9}{4}r$, base$=3r$.

Ex. 265. To find the greatest ellipse that can be inscribed in a semicircle.

Solution. Let r denote the radius of the given semicircle,

a and β the semiaxes of the ellipse,

x and y the co-ordinates of any point of the curve.

Then, if we take the centre of the circle as the origin of rectangular axes, we shall have for the equation of
the ellipse $\quad a^2y^2-2a^2\beta y+\beta^2x^2=0,\quad$ (1)
the circle $\quad y^2+x^2-r^2=0,\quad$ (2)
and $\quad a\beta=$ a *maximum*. $\quad$ (3)

Eliminating x^2 between (1) and (2), we obtain

$$a^2y^2-2a^2\beta y+r^2\beta^2-\beta^2y^2=0, \quad (4)$$

or $\quad (a^2-\beta^2)y^2-2a^2\beta y=-r^2\beta^2, \quad (5)$

$$(a^2-\beta^2)^2y^2-2a^2\beta(a^2-\beta^2)y=-r^2\beta^2(a^2-\beta^2). \quad (6)$$

Comp. $\square\quad (a^2-\beta^2)^2y^2-2a^2\beta(a^2-\beta^2)y+a^4\beta^2=$
$$a^4\beta^2-r^2\beta^2(a^2-\beta^2). \quad (7)$$

If the curves touch each other, the roots of Eq. 7 are equal, and the absolute term is equal to *zero*.

$\therefore a^4\beta^2-r^2\beta^2(a^2-\beta^2)=0, \quad (8)$
$r^2\beta^2=r^2a^2-a^4. \quad (9)$

Multiply both members by a^2, and we have

$$r^2a^2\beta^2=r^2a^4-a^6. \quad (10)$$

The first member of this equation is a maximum, hence the second member must also be a maximum.

$\therefore r^2a^4-a^6=maximum.$

If we put the first differential coefficient equal to zero, we shall have

$a=\frac{1}{3}r\sqrt{6},$
$\beta=\frac{1}{3}r\sqrt{2},$

and $\quad x=y.$

The second differential coefficient becomes *negative*, hence it is a *maximum*.

Ex. 266. In a given semicircle inscribe the greatest

isosceles triangle, having its vertex in the extremity of the diameter, and one of its equal sides coinciding with that diameter.

Ans. An equal side $=1.63$ R.

Ex. 267. Upon a given line (h) as the hypothenuse, to describe a right angled triangle such that the perpendicular added to twice the base shall be a maximum.

Ans. Base $=\dfrac{2h}{\sqrt{5}}$.

Perpendicular $=\dfrac{h}{\sqrt{5}}$.

Ex. 268. Circumscribe about a given parabola the least isosceles triangle.

Ex. 269. Find the radius of the sphere which, placed in a conical glass full of water, shall cause the greatest quantity of water to overflow; the radius of the cone's base being 8, depth of cone $=15$, and slant height $=17$ inches.

Ans. $R = 6.868+$ inches.

173. In the application of the theory of maxima and minima to geometrical investigations, we must be careful not to adopt results which are inconsistent with the geometrical restrictions of the problem; for, when the geometrical conditions of a problem are translated into algebraic language, that language is not necessarily restricted to these conditions; but, while it includes all possible solutions of the problem, it may also furnish others that have no geometrical signification whatever. If, therefore, among such solutions we find any belonging to maxima or minima, they must be rejected.

Ex. 270. Find the value of x that will render the subtangent of the cissoid a maximum.

Ans. $x = \dfrac{3a - a\sqrt{3}}{2}$.

The other value of x, *viz.*, $x = \dfrac{3a + a\sqrt{3}}{2}$, is inadmissible, because it makes y, in the equation of the curve, imaginary.

Maxima and Minima of Functions of two Variables.

174. If u be a function of two variables, x and y, and we put the partial differential coefficients of u regarded as a function of x, and u regarded as a function of y respectively equal to *zero*, and find the values of the variables from these equations, then, if

$$\frac{d^2u}{dx^2} \cdot \frac{d^2u}{dy^2} > \left(\frac{d^2u}{dxdy}\right)^2,$$

the second partial differential coefficients having the same algebraic sign, u will be a maximum when that sign is *negative*, and a minimum when that sign is *positive*.

If the second differential coefficients reduce to *zero* on substituting the values of x and y determined by putting $\frac{du}{dx}=0$ and $\frac{du}{dy}=0$, the function will not be a maximum or minimum.

For, in Art. 64, $\quad u = f(x, y),$
and $\quad\quad\quad\quad\quad u' = f(x+h, y+k);$
then $\quad\quad u'-u = +\frac{du}{dx}h + \frac{d^2u}{dx^2}\frac{h^2}{1.2}+$, etc.,
$$+\frac{du}{dy}k + \frac{d^2u}{dxdy} kh +, \text{ etc.},$$
$$+ \frac{d^2u}{dy^2}\frac{k^2}{1.2}+, \text{ etc.}$$

If, in Art. 64, we had taken
$\quad\quad\quad\quad\quad u = f(x, y),$
and $\quad\quad\quad\quad u'' = f(x-h, y-k),$
we would have had
$$u''-u = -\frac{du}{dx}h + \frac{d^2u}{dx^2}\frac{h^2}{1.2}-, \text{ etc.},$$
$$-\frac{du}{dy}k + \frac{d^2u}{dxdy} kh -, \text{ etc.},$$
$$+ \frac{d^2u}{dy^2}\frac{k^2}{1.2}-, \text{ etc.}$$

If $\frac{du}{dx}$ and $\frac{du}{dy}$ be of finite values, it is obvious that the quantities h and k may be taken so small that the first terms of the second members of $u'-u$ and $u''-u$ may

be greater than the sums of all the terms that follow; in which case one of them will be *positive* and the other *negative*, and therefore u can not be a maximum or a minimum; hence, by reason of the independence of x and y, and therefore of h and k, it follows that
$$\frac{du}{dx}=0, \text{ and } \frac{du}{dy}=0.$$

But, to insure a maximum or a minimum for u, it will be necessary also to have a relation between x and y such that the second term of $u'-u$ or $u''-u$ shall have the same algebraic sign whatever values be attributed to h and k, whether positive or negative.

Now the expression
$$\frac{d^2u}{dx^2}\frac{h^2}{1.2}+\frac{d^2u}{dxdy}kh+\frac{d^2u}{dy^2}\frac{k^2}{1.2}$$
will not change its sign if it consists of the sum of two squares or multiples of two squares. But the expression is evidently equal to

$$\tfrac{1}{2}\frac{d^2u}{dx^2}\left\{\left[h+\frac{\frac{d^2u}{dxdy}}{\frac{d^2u}{dx^2}}k\right]^2 + \left[\frac{\frac{d^2u}{dx^2}\cdot\frac{d^2u}{dy^2}-\left(\frac{d^2u}{dxdy}\right)^2}{\left(\frac{d^2u}{dx^2}\right)^2}k^2\right]\right\},$$

and therefore will not change its algebraic sign whatever values are given to h and k, provided
$$\frac{d^2u}{dx^2}\cdot\frac{d^2u}{dy^2}>\left(\frac{d^2u}{dxdy}\right)^2.$$

The truth of the proposition is therefore manifest.

Examples.

Ex. 271. Let $u=x^4+y^4-4axy^2$, to determine x and y when u is a maximum or minimum.

$$\frac{du}{dx}=4x^3-4ay^2=0;\quad \therefore x^3=ay^2;$$

$$\frac{du}{dy}=4y^3-8axy=0;\quad \therefore y^2=2ax.$$

From which we find $x=\pm a\sqrt{2}$, and $y=a\sqrt[3]{8}$.
Differentiating again, we have
$$\frac{d^2u}{dx^2}=12x^2=24a^2,$$

$$\frac{d^2u}{dy^2} = 12y^2 - 8ax = 16a^2\sqrt{2}.$$

And $\quad \dfrac{d^2u}{dxdy} = -8ay = -8a^2\sqrt[4]{8}$;

$$\therefore \frac{d^2u}{dx^2} \cdot \frac{d^2u}{dy^2} > \frac{d^2u}{dxdy}.$$

Since the algebraic sign of $\dfrac{d^2u}{dx^2}$ and $\dfrac{d^2u}{dy^2}$ is positive, the values of x and y as above give u a minimum.

Ex. 272. Let $u = a\{\sin. x + \sin. y + \sin. (x+y)\}$; find the values of x and y when u is a maximum.

$$\frac{du}{dx} = a\{\cos. x + \cos. (x+y)\} = 0, \quad (1)$$

$$\frac{du}{dy} = a\{\cos. y + \cos. (x+y)\} = 0; \quad (2)$$

$\therefore x = y$.

Substituting y for x in Eqs. (1) or (2), we have

$\cos. x + \cos. 2x = 0$; $\therefore \cos. x + 2\cos.^2 x - 1 = 0$.

Whence $\quad \cos. x = \frac{1}{2}$; $\therefore x = 60° = y$.

And $\quad \dfrac{d^2u}{dx^2} = a\{-\sin. x - \sin. (x+y)\}$

$\qquad = -a\{\sin. 60° + \sin. 120°\}$,

$\qquad = -a\sqrt{3}$;

$$\frac{d^2u}{dy^2} = a\{-\sin. y - \sin. (x+y)\} = -a\sqrt{3}.$$

Hence $\quad \dfrac{d^2u}{dx^2} \cdot \dfrac{d^2u}{dy^2} > \dfrac{d^2u}{dxdy}$;

and since the algebraic sign of $\dfrac{d^2u}{dx^2}$ and $\dfrac{d^2u}{dy^2}$ is negative,

therefore $\quad u = \frac{3}{2}a\sqrt{3} = $ a *maximum*.

Ex. 273. A cistern in the form of a rectangular parallelopipedon is to contain a certain quantity of water; determine its form so that the interior surface shall be a minimum.

Ans. The base must be a square, and the depth = half the length or breadth.

Ex. 274. In a given circle inscribe a triangle whose perimeter shall be a maximum.

Ans. The triangle is equilateral.

Ex. 275. If a square be inscribed in a given circle, a circle in that square, again a square in that circle, and so on continually, prove that the sum of all the inscribed squares is equal to the area of the square circumscribing the given circle.

Ex. 276. If the greatest rectangle be inscribed in an ellipse, the greatest ellipse in that rectangle, again the greatest rectangle in that ellipse, and so on continually, prove that the sum of all the inscribed rectangles is equal to the area of a parallelogram which circumscribes the given ellipse.

Ex. 277. The centres of two spheres are at the extremities of a given straight line, on which a circle is described; find a point in the circumference of the circle from which the greatest portion of spherical surface is visible.

Ex. 278. Of all the ellipses that can be inscribed in a rhombus whose diagonals are $2a$ and $2b$, prove that the greatest is that whose transverse axis is $\dfrac{a}{\sqrt{2}}$ and conjugate axis is $\dfrac{b}{\sqrt{2}}$.

To Change the Independent Variable.

175. If we change an equation between x and y to the form $x=f(y)$, y is called the independent and x the dependent variable.

Let it be required to change the differential coefficient found on the supposition that $x=f(y)$ into another in which both y and x are considered functions of a third variable, t.

Let
$$x=f(y), \quad (1)$$
and
$$y=f'(t). \quad (2)$$
Then, by virtue of these two equations, we must have
$$x=f''(t). \quad (3)$$

Now, if h and k denote the contemporaneous increments of t and y, we shall have from (2) and (3),

$$k=\frac{dy}{dt}h+\frac{d^2y}{dt^2}\frac{h^2}{1\cdot 2}+\frac{d^3y}{dt^3}\frac{h^3}{1\cdot 2\cdot 3}+, \text{etc.}, \quad (4)$$

and
$$x'-x=\frac{dx}{dt}h+\frac{d^2x}{dt^2}\frac{h^2}{1\cdot 2}+\frac{d^3x}{dt^3}\frac{h^3}{1\cdot 2\cdot 3}+, \text{etc.} \quad (5)$$

But from (1) we also have
$$x'-x=\frac{dx}{dy}k+\frac{d^2x}{dy^2}\frac{k^2}{1.2}+\frac{d^3x}{dy^3}\frac{k^3}{1.2.3}+, \text{etc.} \quad (6)$$
Substituting in (6) the value of k in (4), it becomes
$$x'-x=\frac{dx}{dy}\left[\frac{dy}{dt}h+\frac{d^2y}{dt^2}\frac{h^2}{1.2}+, \text{etc.}\right]$$
$$+\frac{d^2x}{dy^2}\frac{1}{1.2}\left[\frac{dy}{dt}h+\frac{d^2y}{dt^2}\frac{h^2}{1.2}+, \text{etc.}\right]^2$$
$$+\frac{d^3x}{dy^3}\frac{1}{1.2.3}\left[\frac{dy}{dt}h+\frac{d^2y}{dt^2}\frac{h^2}{1.2}+, \text{etc.}\right]^3+, \text{etc.} \quad (7)$$

Equating the coefficients of the same powers of h in the two values of $(x'-x)$ in (5) and (7), we have

$$\frac{dx}{dt}=\frac{dx}{dy}\cdot\frac{dy}{dt}; \quad \therefore \frac{dx}{dy}=\frac{\dfrac{dx}{dt}}{\dfrac{dy}{dt}}, \quad (8)$$

$$\frac{d^2x}{dt^2}=\frac{dx}{dy}\cdot\frac{d^2y}{dt^2}+\frac{d^2x}{dy^2}\cdot\frac{dy^2}{dt^2}; \quad (9)$$

$$\therefore \frac{d^2x}{dy^2}=\frac{\dfrac{d^2x}{dt^2}-\dfrac{dx}{dy}\cdot\dfrac{d^2y}{dt^2}}{\dfrac{dy^2}{dt^2}}. \quad (10)$$

Substituting in (10) the value of $\dfrac{dx}{dy}$ found in (8), and reducing, we obtain

$$\frac{d^2x}{dy^2}=\frac{\dfrac{dy}{dt}\cdot\dfrac{d^2x}{dt^2}-\dfrac{dx}{dt}\cdot\dfrac{d^2y}{dt^2}}{\left(\dfrac{dy}{dt}\right)^3}. \quad (11)$$

This process is called *changing the independent variable from y to t.*

ON TANGENT PLANES AND NORMAL LINES TO CURVED SURFACES.

176. To determine the equation of a tangent plane at any point on a curved surface.

Let $\quad u=f(x, y, z)=0$

be the equation of the surface, and designate the co-ordinates of the point at which the plane is to be tangent by x', y', z'.

Then, if we suppose two planes to be passed through the given point respectively parallel to the co-ordinate planes xz and yz, they will intersect the surface in two curves.

If a tangent line be drawn to each of these curves at the given point of intersection, the plane of these tangents will be tangent to the surface at the given point.

The equations of these tangent lines are

$$z-z'=\frac{dz'}{dx'}(x-x'), \qquad (1)$$

$$y=y', \qquad (2)$$

$$z-z'=\frac{dz'}{dy'}(y-y'), \qquad (3)$$

$$x=x'. \qquad (4)$$

These equations represent the projections of the tangent lines on the co-ordinate planes xz and yz.

Now, since the traces of the plane through these two lines upon the planes xz and yz are parallel to the lines themselves, therefore $\frac{dz'}{dx'}$, the tangent of the angle which the projection of the first tangent on the co-ordinate plane xz makes with the axis of x, must be the same in the trace as in the line. In like manner, $\frac{dz'}{dy'}$, the tangent of the angle which the projection of the second tangent on the co-ordinate plane yz makes with the axis of y, must be the same in the trace as in the line.

The equation of a plane passing through a single point (Anal. Geom., Art. 150) is

$$z-z'=A(x-x')+B(y-y'). \qquad (5)$$

Substituting for A and B their equivalents $\frac{dz'}{dx'}$ and $\frac{dz'}{dy'}$, we have

$$z-z'=(x-x')\frac{dz'}{dx'}+(y-y')\frac{dz'}{dy'}. \qquad (6)$$

But in the equation

$$u=f(x, y, z)=0,$$

the expressions for the total differential coefficients derived from u considered as a function, first of x, and then of y, are

$$\frac{du}{dx} = \frac{du}{dx} + \frac{du}{dz} \cdot \frac{dz'}{dx'} = 0,$$

$$\frac{du}{dy} = \frac{du}{dy} + \frac{du}{dz} \cdot \frac{dz'}{dy'} = 0;$$

From which we obtain

$$\frac{dz'}{dx'} = -\frac{\dfrac{du}{dx}}{\dfrac{du}{dz}},$$

$$\frac{dz'}{dy'} = -\frac{\dfrac{du}{dy}}{\dfrac{du}{dz}}.$$

By substituting these in (6) and reducing, we have

$$(z-z')\frac{du}{dz} + (x-x')\frac{du}{dx} + (y-y')\frac{du}{dy} = 0, \qquad (7)$$

which is the equation of a tangent plane to a curved surface at a point whose co-ordinates are x', y', z'.

177. A normal line to the surface is perpendicular to the tangent at the point of tangency, therefore its equations will be of the form

$$x - x' = \frac{\dfrac{du}{dx}}{\dfrac{du}{dz}}(z-z'),$$

$$y - y' = \frac{\dfrac{du}{dy}}{\dfrac{du}{dz}}(z-z').$$

178. *To find the radius of curvature of a curve of double curvature, the arc of the curve being considered as the independent variable.*

Let the angles which a normal to the curve at any point makes with the co-ordinate planes be denoted by V', V'', and V'''; and the angles which a normal at any

point near to the former makes with the same planes be denoted by V_1, V_2, V_3; so that
$$V' = \cos.^{-1} x_{,}, \quad V'' = \cos.^{-1} y_{,}, \quad V''' = \cos.^{-1} z_{,};$$
$$V_1 = \cos.^{-1}(x_{,}+h),$$
$$V_2 = \cos.^{-1}(y_{,}+k),$$
$$V_3 = \cos.^{-1}(z_{,}+l).$$

Then, if V denotes the angle between the normals, we shall have, by Anal. Geom., Art. 158,
$$\cos. V = x_{,}(x_{,}+h) + y_{,}(y_{,}+k) + z_{,}(z_{,}+l),$$
$$= x_{,}^2 + x_{,}h + y_{,}^2 + y_{,}k + z_{,}^2 + z_{,}l.$$

But, by Anal. Geom., Art. 157,
$$x_{,}^2 + y_{,}^2 + z_{,}^2 = 1;$$
$$\therefore \cos. V = 1 + x_{,}h + y_{,}k + z_{,}l,$$
or $\quad 1 - \cos. V = -(x_{,}h + y_{,}k + z_{,}l).$ \quad (A)

Again: $(x_{,}+h)^2 + (y_{,}+k)^2 + (z_{,}+l)^2 = 1.$
Actually squaring, and reducing as above, we have
$$x_{,}h + y_{,}k + z_{,}l = -\tfrac{1}{2}(h^2 + k^2 + l^2). \quad (B)$$
But $\quad 1 - \cos. V = 2 \sin.^2 \tfrac{1}{2} V,$
$$= \tfrac{1}{2} V^2,$$
when the arc is indefinitely diminished.

Substituting in (A) the value of $1 - \cos. V$, and that of $x_{,}h + y_{,}k + z_{,}l$, we obtain
$$V^2 = h^2 + k^2 + l^2.$$

The radius of curvature, which we will denote by ρ, is
$$\rho = \frac{ds}{V} = \frac{ds}{\sqrt{h^2 + k^2 + l^2}}.$$

But, by Art. 91,
$$x_{,} = \frac{dx}{ds}, \quad y_{,} = \frac{dy}{ds}, \text{ and } z_{,} = \frac{dz}{ds}.$$

Moreover, when the normals become consecutive,
$$h^2 = \left(\frac{d^2x}{ds}\right)^2, \quad k^2 = \left(\frac{d^2y}{ds}\right)^2, \quad l^2 = \left(\frac{d^2z}{ds}\right)^2$$

$$\therefore \rho = \frac{ds}{\sqrt{\left(\frac{d^2x}{ds}\right)^2 + \left(\frac{d^2y}{ds}\right)^2 + \left(\frac{d^2z}{ds}\right)^2}},$$

or
$$\rho = \frac{ds^2}{\sqrt{(d^2x)^2 + (d^2y)^2 + (d^2z)^2}}.$$

ELEMENTS OF THE
INTEGRAL CALCULUS.

179. The Integral Calculus explains the method of finding the function from which a given differential has been deduced.

In the Differential Calculus we have a system of rules by means of which we can deduce from any given function a second function, which is called the differential of the former; but in the Integral Calculus the rules are not so direct; they are deduced by reversing the process by which we obtained the differential from the function.

The terms *integral* and *integration* are taken from the *infinitesimal calculus*. In the infancy of the science, differentials were considered as *infinitely small quantities;* hence the original functions from which these differentials were deduced were called the sums of the infinitely small elements; and the process by which these original functions were obtained from their differentials was called the *summation* or *integration* of the indefinitely small component parts, and the operation was expressed by the character $\int$ prefixed to the differential. Thus, the integral of $x^n dx$ was written

$$\int x^n dx.$$

This character is still retained by most writers upon the subject.

We shall first take up those differentials which are functions of a single variable, and show how they may be integrated.

180. Let us find the integral of
$$x^{n-1} dx.$$
We have seen (Art. 48) that
$$d \cdot x^n = n x^{n-1} dx;$$
$$\therefore x^{n-1} dx = \frac{d \cdot x^n}{n}.$$

Now as the numerator of the fraction in the second member is only the indication of a differential, its integral is evidently the function x^n. Therefore

$$\int x^{n-1} dx = \frac{x^n}{n},$$

from which we derive the following

RULE.

To integrate a monomial of the form $x^{n-1} dx$, add unity to the exponent of the variable, divide by the exponent so increased, and drop the differential of the variable.

Examples.

Ex. 1. Find the integral of $\dfrac{dx}{x^{\frac{2}{3}}}$.

Bringing the denominator into the numerator by changing the sign of its exponent, we have

$$\frac{dx}{x^{\frac{2}{3}}} = x^{-\frac{2}{3}} dx.$$

Then, by the rule above given,

$$\int \frac{dx}{x^{\frac{2}{3}}} = \int x^{-\frac{2}{3}} dx = \frac{x^{-\frac{2}{3}+1}}{-\frac{2}{3}+1} = 3x^{\frac{1}{3}}.$$

181. In (Art. 41) it was shown that the differential of a variable function and the differential of the same function increased or diminished by a constant quantity were both equal; hence the same differential may answer for several integral functions differing from each other in the value of the constant term.

In integrating a differential, therefore, we must always add a constant quantity to the first integral that we obtain, and then find such a value for this constant as will characterize the particular integral required.

The integral first obtained is called the *indefinite integral*, but after it has been corrected by the method just named it is called the *definite integral*.

$$\therefore \int \frac{dx}{x^{\frac{2}{3}}} = 3x^{\frac{1}{3}} + c, \ c \text{ being the correction.}$$

Ex. 2. Integrate $ax^6 dx$.

L

182. By (Art. 38), the differential of a constant multiplied by a variable function is equal to the constant into the differential of the function. Hence we may conclude that *the integral of a constant into a variable function is equal to the constant into the integral of the differential.* Therefore, *if the expression to be integrated has a constant factor, it may be placed in front of the sign of the integral.*

Hence
$$\int ax^6 dx = a \int x^6 dx,$$
$$= a \cdot \frac{x^7}{7},$$
$$= \tfrac{1}{7}ax^7 + c.$$

183. There is one case in which the rule (Art. 180) does not apply: that is, when the exponent of the variable is equal to *minus unity.* Thus, by the rule,
$$\int x^{-1} dx = \frac{x^0}{0} = \frac{1}{0} = \infty,$$
which is not true. We can, however, take a factor from the numerator and place it in the denominator by changing the sign of its exponent.
$$\therefore x^{-1} dx = \frac{dx}{x}.$$

Now we know from (Art. 69) that
$$d \cdot \log x = \frac{dx}{x},$$
taken in the Naperian system;
$$\therefore \int \frac{dx}{x} = \log x + c.$$

The integral of every fraction whose numerator is the differential of its denominator is the Naperian logarithm of the denominator.

Integrate the following:

Exs. (3.) $\int x^{\frac{1}{3}} dx.$ (4.) $\int \frac{dx}{\sqrt{x}}.$ (5.) $\int \frac{adx}{bx^{\frac{1}{6}}}.$

(6.) $\int 7x^6 dx.$ (7.) $\int x^9 dx.$ (8.) $\int 5x^3 dx.$ (9.) $\int \tfrac{7}{9}x^{\frac{5}{3}} dx.$

(10.) $\int \frac{6dx}{x^3}.$ (11.) $\int \tfrac{1}{3}x^2 dx.$ (12.) $\int \frac{dx}{x^3}.$ (13.) $\int ax^2 dx.$

(14.) $\int ax^3 dx.$ (15.) $\int \frac{adx}{x}.$ (16.) $\int \frac{adx}{x^2}.$ (17.) $\int \frac{adx}{x^3}.$

INTEGRALS. 243

(18.) $\int \dfrac{adx}{x^4}$.   (19.) $\int \dfrac{adx}{x^{\frac{1}{3}}}$.   (20.) $\int \dfrac{adx}{x^{\frac{2}{3}}}$.   (21.) $\int \dfrac{adx}{x^{\frac{3}{4}}}$.

(22.) $\int \dfrac{dx}{5x^2}$.

184. Since the differential of the sum or difference of any number of functions is equal to the sum or difference of their differentials, it follows that *the integral of the sum or difference of any number of differentials must be equal to the sum or difference of their integrals taken separately.* Thus, if

Ex. 23. $du = 2a^2 x dx - 6ax^2 dx + dx$,
then $u = 2a^2 \int x dx - 6a \int x^2 dx + \int dx$,
  $u = a^2 x^2 - 2ax^3 + x + c$.

Ex. 24. $du = x dx - \dfrac{x^2 dx}{c} + bx^{\frac{2}{3}} dx - \dfrac{adx}{x^{\frac{1}{2}}}$,

$$u = \int x dx - \dfrac{1}{c}\int x^2 dx + b \int x^{\frac{2}{3}} dx - a \int \dfrac{dx}{\sqrt{x}}.$$

185. If we have a polynomial of the form
$$(a + bx + cx^2 +, \text{etc.})^n dx,$$
where n is a *positive whole number*, we can integrate it by first raising it to the power indicated by n, multiplying each term by dx, and then integrating each term separately. Thus, if we have

$\int (a + bx + cx^2)^2 dx = a^2 \int dx + 2ab \int x dx + (2ac + b^2) \int x^2 dx$
$\qquad\qquad + 2bc \int x^3 dx + c^2 \int x^4 dx;$

and integrating each term separately, we have

$\int (a + bx + cx^2)^2 dx = a^2 x + abx^2 + \frac{1}{3}(2ac + b^2)x^3$
$\qquad\qquad + \frac{1}{2}bcx^4 + \frac{1}{5}c^2 x^5 + C.$

186. Many expressions may be transformed into monomials by the introduction of an auxiliary variable, and then integrated by the first rule.

Let $du = (a + bx^n)^p m x^{n-1} dx$.
Put $a + bx^n = v$;
then $nbx^{n-1} = dv$;
  $\therefore x^{n-1} = \dfrac{dv}{nb}$.

Substituting these in the given expression, we have
$$du = \dfrac{mv^p dv}{nb};$$

$$\therefore u = \frac{m}{nb}\int v^p dv = \frac{m}{nb} \cdot \frac{v^{p+1}}{p+1},$$

and, replacing the value of v,

$$u = \frac{m(a+bx^n)^{p+1}}{nb(p+1)} + c.$$

Hence, to integrate a binomial differential when the exponent of the variable without the parenthesis is less by unity than the exponent of the variable within, we derive the following

Rule.

Multiply the binomial, with its primitive exponent increased by unity, by the constant factor, if there be one; then divide this result by the continued product of the new exponent, the coefficient, and the exponent of the variable within the parenthesis.

Examples.

Ex. 25. Integrate $du = (a^2+6x^3)^{-\frac{1}{2}}x^2 dx.$

Ans. $u = \frac{1}{9}(a^2+6x^3)^{\frac{1}{2}} + c.$

Ex. 26. If $du = (a-x)^{-5}b\,dx,$

then $u = \dfrac{b}{4(a-x)^4} + c.$

Ex. 27. If $du = (a+bx^2)^{\frac{3}{2}}x\,dx,$

then $u = \dfrac{(a+bx^2)^{\frac{5}{2}}}{5b}.$

Ex. 28. If $du = ax^8 dx (a^9+6x^9)^{-\frac{1}{2}},$

then $u = \dfrac{a(a^9+6x^9)^{\frac{1}{2}}}{27} + c.$

Ex. 29. If $du = 3(2-3x^5)^{-\frac{1}{2}}x^4 dx,$

then $u = -\frac{2}{5}(2-3x^5)^{\frac{1}{2}} + c.$

Ex. 30. If $du = (1-x^2)^{-2} 4x\,dx,$

then $u = \dfrac{2}{1-x^2} + c.$

187. In a manner entirely similar we may integrate the following differentials.

INTEGRALS OF LOGARITHMIC FUNCTIONS. 245

If $$du = \frac{x^{n-1}dx}{a+x^n},$$
we may put $a+x^n = v$;
then $nx^{n-1}dx = dv,$
and $x^{n-1}dx = \frac{dv}{n}.$

Substituting these, we have
$$du = \frac{dv}{n(v+x^n)};$$
$$\therefore u = \frac{1}{n}\int \frac{dv}{v}.$$

But $$\int \frac{dv}{v} = \log. v = \log. (a+x^n);$$
$$\therefore u = \frac{1}{n} \cdot \log. (a+x^n) + c.$$

Ex. 31. Let $$du = \frac{(3x^2+2x+1)dx}{x^3+x^2+x+1}.$$
Put $x^3+x^2+x+1 = v$;
then $(3x^2+2x+1)dx = dv,$
and $$\int du = \int \frac{dv}{v} = \log. v = \log. (x^3+x^2+x+1) + c.$$

We see, then, that when the numerator of an expression is equal to the differential of the denominator, *its integral will be the Naperian logarithm of the denominator.*

Again: let $$du = \frac{p(a+2bx)dx}{q(ax+bx^2)};$$
then $$u = \frac{p}{q} \int \frac{(a+2bx)dx}{ax+bx^2}.$$

By the foregoing we find
$$\int \frac{(a+2bx)dx}{ax+bx^2} = \log. (ax+bx^2) + c;$$
$$\therefore u = \frac{p}{q} \cdot \log. (ax+bx^2) + c.$$

188. Every expression of the form
$$du = ax^m(b+cx)^n dx$$
can be integrated when *either m or n is a positive whole number.*

For, when n is positive, we can raise the binomial to

the power indicated by n, and then multiply every term by $ax^m dx$, and integrate each term separately.

And when n is fractional or negative, m being positive and entire,

put $\qquad b + cx = v;$

then $\qquad x = \dfrac{v-b}{c},$

and $\qquad dx = \dfrac{dv}{c};$

$$\therefore du = \frac{a}{c}\left(\frac{v-b}{c}\right)^m v^n dv;$$

$$\therefore u = \frac{a}{c}\int \left(\frac{v-b}{c}\right)^m v^n dv.$$

Ex. 32. Let $\quad du = 2x dx (1-3x)^{-\frac{1}{2}}.$

Put $\qquad 1 - 3x = v;$

then $\qquad x = \dfrac{1-v}{3},$

and $\qquad dx = -\tfrac{1}{3} dv;$

$$\therefore du = -\tfrac{2}{9}(1-v) v^{-\frac{1}{2}} dv,$$

$$= -\tfrac{2}{9} v^{-\frac{1}{2}} dv + \tfrac{2}{9} v^{\frac{1}{2}} dv;$$

$$\therefore u = -\tfrac{4}{9} v^{\frac{1}{2}} + \tfrac{4}{27} v^{\frac{3}{2}},$$

or $\qquad = \tfrac{4}{27}(1-3x)^{\frac{3}{2}} - \tfrac{4}{9}(1-3x)^{\frac{1}{2}} + c.$

Ex. 33. Integrate $du = bx^2(a-x)^{\frac{1}{2}} dx.$

Ans. $u = \tfrac{4}{5}ab(a-x)^{\frac{5}{2}} - \tfrac{2}{3}ab(a-x)^{\frac{3}{2}} - \tfrac{2}{7}b(a-x)^{\frac{7}{2}} + c.$

Ex. 34. If $\qquad du = \dfrac{x dx}{\sqrt{a^2 + x^2}},$

$$u = \sqrt{a^2 + x^2} + C.$$

Ex. 35. If $\qquad du = \dfrac{x dx}{\sqrt{a^2 - x^2}},$

$$u = -\sqrt{a^2 - x^2} + C.$$

Ex. 36. If $\qquad du = \dfrac{x dx}{(a^2 + x^2)^{\frac{3}{2}}},$

$$u = -(a^2 + x^2)^{-\frac{1}{2}} + C.$$

INTEGRATION OF CIRCULAR FUNCTIONS. 247

Ex. 37. If $\quad du = \dfrac{xdx}{(a^2-x^2)^{\frac{3}{2}}},$

$\quad u = (a^2-x^2)^{-\frac{1}{2}} + C.$

Ex. 38. If $\quad du = (a-x)(2ax-x^2)^{\frac{1}{2}}dx.$
Put $\quad 2ax - x^2 = z;\ \therefore (a-x)dx = \tfrac{1}{2}dz,$
and $\quad (a-x)(2ax-x^2)^{\frac{1}{2}} = \tfrac{1}{2}z^{\frac{1}{2}}dz;$

$\therefore \int (a-x)(2ax-x^2)^{\frac{1}{2}}dx = \dfrac{(2ax-x^2)^{\frac{3}{2}}}{3} + C.$

Ex. 39. If $\quad du = \dfrac{(a+x)dx}{\sqrt{2ax+x^2}},$

$\quad u = \sqrt{2ax+x^2} + C.$

Ex. 40. If $\quad du = \dfrac{\sqrt{2ax-x^2}}{\sqrt{x}} \cdot dx,$

$\quad u = -\tfrac{2}{3}(2a-x)^{\frac{3}{2}} + C.$

Ex. 41. If $\quad du = (a+bx+cx^2)^{\frac{2}{3}}(b+2cx)dx,$

$\quad u = \tfrac{3}{5}(a+bx+cx^2)^{\frac{5}{3}} + C.$

Ex. 42. If $\quad du = \dfrac{5x^3 dx}{3x^4+7},$

$\quad u = \tfrac{5}{12} \log.(3x^4+7) + C.$

INTEGRATION OF THE DIFFERENTIALS OF CIRCULAR FUNCTIONS.

189. If x denotes the arc whose sign is u, we have, by (Art. 84),

$$dx = \dfrac{du}{\sqrt{1-u^2}};$$

$$\therefore x = \int \dfrac{du}{\sqrt{1-u^2}} = \sin.^{-1} u + C.$$

If we let $\quad u = 0,$ then x, or the arc $= 0 + C,$
and if $\quad u = 1,$ then x, or the arc $= \tfrac{1}{2}\pi + C.$

Now, taking the difference between these two, we shall eliminate the constant C; then

$$\int_0^1 \dfrac{du}{\sqrt{1-u^2}} = \tfrac{1}{2}\pi.$$

The expression $\int_0 \dfrac{du}{\sqrt{1-u^2}}$ is called *the integral taken between the limits:* $u=0$, and $u=1$.

Ex. 43. $\int dx = \int -\dfrac{du}{\sqrt{1-u^2}} = \cos.^{-1} u + C.$

Ex. 44. $\int \dfrac{du}{\sqrt{2u-u^2}} = \text{versin.}^{-1} u + C.$

Ex. 45. $\int \dfrac{du}{1+u^2} = \tan.^{-1} u + C.$

Ex. 46. $\int -\dfrac{du}{1+u^2} = \cot.^{-1} u + C.$

Ex. 47. $\int \dfrac{du}{u\sqrt{u^2-1}} = \sec.^{-1} u + C.$

Ex. 48. $\int -\dfrac{du}{u\sqrt{u^2-1}} = \operatorname{cosec.}^{-1} u + C.$

Ex. 49. Integrate $\dfrac{du}{\sqrt{a^2-u^2}} = \dfrac{du}{a\sqrt{1-\dfrac{u^2}{a^2}}}.$

Put $\dfrac{u^2}{a^2} = v^2;$

then $\dfrac{u}{a} = v,$ and $\dfrac{du}{a} = dv.$

Substituting these in the given expression, we shall have
$$\int \dfrac{dv}{\sqrt{1-v^2}} = \sin.^{-1} v = \sin.^{-1} \dfrac{u}{a} + C.$$

Ex. 50. Integrate $-\dfrac{du}{\sqrt{a^2-u^2}}.$

$$= \cos.^{-1} \dfrac{u}{a} + C.$$

Ex. 51. Integrate $\dfrac{dx}{a^2+x^2}.$

$$\int \dfrac{dx}{a^2+x^2} = \int \dfrac{dx}{a^2\left(1+\dfrac{x^2}{a^2}\right)}.$$

INTEGRATION OF CIRCULAR FUNCTIONS. 249

Put $$\frac{x^2}{a^2}=v^2;$$

$$\therefore \frac{x}{a}=v, \text{ and } dx=adv.$$

Substituting,

$$\int \frac{dx}{a^2+x^2}=\frac{1}{a}\int \frac{dv}{1+v^2}=\frac{1}{a}\tan.^{-1}\frac{x}{a}+C.$$

Ex. 52. Integrate $\dfrac{dx}{\sqrt{2ax-x^2}}=\dfrac{dx}{a\sqrt{\dfrac{2x}{a}-\dfrac{x^2}{a^2}}}.$

Put $$\frac{x^2}{a^2}=v^2;$$

then $$\frac{x}{a}=v, \text{ and } dx=adv.$$

Substituting,

$$\int \frac{dx}{\sqrt{2ax-x^2}}=\int \frac{dv}{\sqrt{2v-v^2}}=\text{versin.}^{-1} v,$$

$$=\text{versin.}\frac{x}{a}+C.$$

In these integrals the radius of the circle is unity; but as it is frequently desirable to integrate differentials in which the radius is r instead of unity, we shall exhibit a few of those which occur most frequently to that radius. Thus,

Ex. 53. $\displaystyle\int \frac{Rdu}{\sqrt{R^2-u^2}}=\sin.^{-1} u+C.$

Ex. 54. $\displaystyle\int -\frac{Rdu}{\sqrt{R^2-u^2}}=\cos.^{-1} u+C.$

Ex. 55. $\displaystyle\int \frac{Rdu}{\sqrt{2Ru-u^2}}=\text{versin.}^{-1} u+C.$

Ex. 56. $\displaystyle\int \frac{R^2du}{R^2+u^2}=\tan.^{-1} u+C.$

Ex. 57. Integrate $du=\sin. x \cos. x . dx.$
Put $\sin. x=v;$
then, by differentiating, $\cos. xdx=dv.$

Substituting for sin. x and cos. xdx their values, we have

or
$$du = v^{n-1}dv,$$
$$u = \frac{1}{n}v^n,$$
$$= \frac{1}{n}\sin^n x + C.$$

Ex. 58. If $du = \sin^2 x \cos x \, dx,$
then $u = \frac{1}{3}\sin^3 x + C.$

Ex. 59. If $du = \sin^{\frac{1}{2}} x \cos x \, dx,$
then $u = \frac{2}{3}\sin^{\frac{3}{2}} x + C.$

Ex. 60. Integrate $du = (\tan^n \theta + \tan^{n+2} \theta)d\theta,$
$$= \tan^n \theta(1 + \tan^2 \theta)d\theta.$$
Put $\tan \theta = v;$
then, by differentiating, we have
$$(1 + \tan^2 \theta)d\theta = dv,$$
and substituting, $du = v^n dv;$
$$\therefore u = \frac{1}{n+1}\tan^{n+1}\theta + C.$$

Ex. 61. If $du = (\tan^5 \theta + \tan^7 \theta)d\theta,$
then $u = \frac{1}{6}\tan^6 \theta + C.$

Ex. 62. Integrate $du = \sin 2x \, dx.$
By (Trig., Prob. XVI.), $\sin 2x = 2 \sin x \cos x;$
$$\therefore du = 2 \sin x \cos x \, dx.$$
$$u = \sin^2 x + C.$$

Ex. 63. If $du = (\cot^3 \theta + \cot^5 \theta)d\theta,$
$$u = -\frac{1}{4}\cot^4 \theta + C.$$

Ex. 64. Integrate $\dfrac{dx}{\sqrt{x^2 \pm a^2}}.$

Put $\sqrt{x^2 \pm a^2} = v;$
then $x^2 \pm a^2 = v^2;$
hence, by differentiating, $x \, dx = v \, dv.$
Putting this into a proportion,
$$x : v :: dv : dx,$$
or, by composition, $x + v : v :: dx + dv : dx;$
$$\therefore \frac{dx + dv}{x + v} = \frac{dx}{v} = \frac{dx}{\sqrt{x^2 \pm a^2}};$$
$$\therefore \int \frac{dx}{\sqrt{x^2 \pm a^2}} = \log(x + v),$$
or, substituting for v its value,
$$= \log(x + \sqrt{x^2 \pm a^2}).$$

INTEGRATION OF CIRCULAR FUNCTIONS. 251

Ex. 65. Integrate $\dfrac{dx}{\sqrt{x^2 \pm 2ax}}$.

Put $\sqrt{x^2 \pm 2ax} = v$;
then $x^2 \pm 2ax = v^2$,
$$x^2 \pm 2ax + a^2 = v^2 + a^2,$$
$$x \pm a = \sqrt{v^2 + a^2};$$
$$\therefore dx = \dfrac{v\,dv}{\sqrt{v^2 + a^2}}.$$

Hence $\dfrac{dx}{\sqrt{x^2 \pm 2ax}} = \dfrac{dv}{\sqrt{v^2 + a^2}}$,

and, by Ex. (64), $\displaystyle\int \dfrac{dv}{\sqrt{v^2 + a^2}} = \log.(v + \sqrt{v^2 + a^2})$;

$$\therefore \int \dfrac{dx}{\sqrt{x^2 \pm 2ax}} = \log.\{\sqrt{x^2 \pm 2ax} + x \pm a\} + C.$$

Ex. 66. Integrate $\dfrac{2dx}{4 + x^2}$.

$$\int \dfrac{2dx}{4 + x^2} = \int \dfrac{\frac{1}{2}dx}{1 + \dfrac{x^2}{4}}.$$

Put $\dfrac{x^2}{4} = v^2$;

$$\therefore \dfrac{x}{2} = v, \text{ and } \tfrac{1}{2}dx = dv.$$

Substitute, and we have
$$\int \dfrac{dv}{1 + v^2} = \tan.^{-1} v + C,$$
$$= \tan.^{-1} \dfrac{x}{2} + C.$$

Ex. 67. If $du = \dfrac{3dx}{1 + 4x^2}$,
$$u = \tfrac{3}{2} \tan.^{-1} 2x + C.$$

Ex. 68. If $du = \dfrac{x\,dx}{\sqrt{a - bx^4}}$,
$$u = \dfrac{1}{2\sqrt{b}} \sin.^{-1} x^2 \sqrt{\dfrac{b}{a}} + C.$$

Ex. 69. If $du = \dfrac{x^{\frac{1}{2}}dx}{a+bx^3}$,

$$u = \dfrac{2}{3\sqrt{ab}} \tan^{-1} x^{\frac{3}{2}} \sqrt{\dfrac{b}{a}} + C.$$

Ex. 70. If $du = \dfrac{x^{n-1}dx}{\sqrt{a-bx^{2n}}}$,

$$u = \dfrac{1}{n\sqrt{b}} \sin^{-1} x^n \sqrt{\dfrac{b}{a}} + C.$$

Ex. 71. If $du = \dfrac{m\,dx}{\sqrt{1-m^2x^2}}$,

$$u = \sin^{-1} mx + C.$$

Ex. 72. If $du = \dfrac{2x+1}{x^2+x+1} \cdot dx$,

$$u = \log.(x^2+x+1) + C.$$

INTEGRATION BY SERIES.

190. Every expression of the form
$$f(x)\,dx$$
that can be developed in the powers of x may be integrated by a series. For if we suppose that
$$f(x) = A_0 + A_1x + A_2x^2 + A_3x^3 +, \text{ etc.},$$
then $f(x)dx = A_0 dx + A_1 x\,dx + A_2 x^2 dx + A_3 x^3 dx +$, etc.;
$$\therefore \int f(x)dx = A_0 x + \tfrac{1}{2}A_1 x^2 + \tfrac{1}{3}A_2 x^3 + \tfrac{1}{4}A_3 x^4 +, \text{ etc.}$$

We have simply to develop the function of x into a series, multiply each term of the series by dx, and integrate each term separately.

Ex. 73. Let us integrate $\dfrac{dx}{\sqrt{x^2-1}} = dx(x^2-1)^{-\frac{1}{2}}$.

If we develop $(x^2-1)^{-\frac{1}{2}}$, we shall have
$$(x^2-1)^{-\frac{1}{2}} = x^{-1} + \tfrac{1}{2}x^{-3} + \dfrac{1 \cdot 3}{2 \cdot 4}x^{-5} + \dfrac{1 \cdot 3 \cdot 5}{2 \cdot 4 \cdot 6}x^{-7} +, \text{ etc.};$$
$$\therefore \dfrac{dx}{\sqrt{x^2-1}} = x^{-1}dx + \tfrac{1}{2}x^{-3}dx + \dfrac{1 \cdot 3}{2 \cdot 4}x^{-5}dx + \dfrac{1 \cdot 3 \cdot 5}{2 \cdot 4 \cdot 6}x^{-7}dx +,$$
etc.; therefore
$$\int \dfrac{dx}{\sqrt{x^2-1}} = \log. x - \dfrac{1}{4x^2} - \dfrac{1 \cdot 3}{32x^4} - \dfrac{1 \cdot 3 \cdot 5}{2 \cdot 4 \cdot 6 \cdot 6x^6} -, \text{ etc., } + C.$$

INTEGRATION OF DIFFERENTIAL BINOMIALS.

191. Every binomial differential can be represented under the general form

$$x^{m-1}dx(a+bx^n)^{\frac{p}{q}},$$

in which m and n are whole numbers, and n positive.

For, if m and n were fractional, and the binomial of the form

$$x^{\frac{1}{3}}dx(a+bx^{\frac{1}{2}})^{\frac{p}{q}},$$

we might substitute for x another variable whose exponent should be equal to the least common multiple of the denominators of the exponents of x. We should then have an expression in which the exponents are whole numbers.

Thus, put $\qquad x=v^6;$
then $\qquad dx=6v^5 dv,$
and $\qquad x^{\frac{1}{3}}dx(a+bx^{\frac{1}{2}})^{\frac{p}{q}}=6v^7 dv(a+bv^3)^{\frac{p}{q}}.$

Again: if n be negative, or

$$x^{m-1}dx(a+bx^{-n})^{\frac{p}{q}},$$

we may put $\qquad x=\dfrac{1}{v};$
then $\qquad dx=-v^{-2}dv;$
and substituting, we have

$$x^{m-1}dx(a+bx^{-n})^{\frac{p}{q}}=-v^{-m-1}dv(a+bv^n)^{\frac{p}{q}},$$

an expression in which the exponents of the variable are whole numbers.

If the variable x is in both terms of the binomial, as

$$x^{m-1}dx(ax^r+bx^n)^{\frac{p}{q}},$$

we may take x^r from under the vinculum, and we shall have

$$x^{m+\frac{pr}{q}-1}dx(a+bx^{n-r})^{\frac{p}{q}},$$

where only one of the terms within the parenthesis contains the variable.

192. Suppose, then, that it be required to find the integral of

$$du=x^{m-1}dx(a+bx^n)^{\frac{p}{q}}.$$

Put
$$a+bx^n=v^q;$$
then
$$bx^n=v^q-a,$$
and
$$b^{\frac{m}{n}}x^m=(v^q-a)^{\frac{m}{n}};$$
$$\therefore b^{\frac{m}{n}}x^{m-1}dx=\frac{q}{n}v^{q-1}dv(v^q-a)^{\frac{m}{n}-1};$$
$$\therefore du=\frac{q}{n(b)^{\frac{m}{n}}}v^{p+q-1}dv(v^q-a)^{\frac{m}{n}-1},$$

an expression which is easily integrated when $\frac{m}{n}$ is a whole number.

Hence *every binomial differential has an exact integral when the exponent of the variable without the parenthesis increased by unity is exactly divisible by the exponent of the variable within.*

193. Again: let us make
$$a+bx^n=v^qx^n;$$
then
$$x^n=a(v^p-b)^{-1};$$
$$\therefore x^m=a^{\frac{m}{n}}(v^q-b)^{-\frac{m}{n}};$$
$$\therefore x^{m-1}dx=-\frac{qa^{\frac{m}{n}}}{n}\cdot\frac{v^{q-1}dv}{(v^q-b)^{\frac{m}{n}+1}},$$
and
$$(a+bx^n)^{\frac{p}{q}}=(v^qx^n)^{\frac{p}{q}}=v^pa^{\frac{p}{q}}(v^q-b)^{-\frac{p}{q}};$$
$$\therefore du=x^{m-1}dx(a+bx^n)^{\frac{p}{q}}=-\frac{qa^{\frac{m}{n}+\frac{p}{q}}}{n}\cdot\frac{v^{p+q-1}dv}{(v^q-b)^{\frac{m}{n}+\frac{p}{q}+1}},$$

which can be integrated when $\frac{m}{n}+\frac{p}{q}$ is a whole number.

Hence *every binomial differential can be integrated when the exponent of the variable without the parenthesis, increased by unity and divided by the exponent of the variable within, and the quotient, added to the exponent of the parenthesis, is an entire number, or zero.*

Examples.

Ex. 74. Integrate $du=xdx(a+bx^2)^3$.

$$\text{Ans. } u=\frac{(a+bx^2)^4}{8b}+C.$$

INTEGRATION OF DIFFERENTIAL BINOMIALS. 255

Ex. 75. Integrate $du = \dfrac{2xdx}{(x^2+1)^2}.$

$$\text{Ans. } u = -\dfrac{1}{x^2+1}.$$

Ex. 76. Integrate $du = \dfrac{dx}{x^4(a^2-x^2)^{\frac{1}{2}}}.$

$$\dfrac{dx}{x^4(a^2-x^2)} = x^{-4}dx(a^2-x^2)^{-\frac{1}{2}}.$$

Put $\sqrt{a^2-x^2} = vx;$

$\therefore a^2-x^2 = v^2x^2,$ and $x^2 = \dfrac{a^2}{1+v^2};$

$$\therefore x = \dfrac{a}{\sqrt{1+v^2}},$$

and $dx = -\dfrac{avdv}{(1+v^2)^{\frac{3}{2}}}.$

Again: $\dfrac{1}{x^2} = \dfrac{1+v^2}{a^2},$ and $\dfrac{1}{x^4} = \dfrac{(1+v^2)^2}{a^4};$

$$\therefore x^{-4} = \dfrac{(1+v^2)^2}{a^4}.$$

Substituting these in the original expression, we have

$$x^{-4}dx(a^2-x^2)^{-\frac{1}{2}} = -\dfrac{(1+v^2)^2}{a^4} \cdot \dfrac{avdv}{(1+v^2)^{\frac{3}{2}}} \cdot \dfrac{1}{vx}.$$

But $\dfrac{1}{x} = \dfrac{\sqrt{1+v^2}}{a};$

$$\therefore \dfrac{1}{vx} = \dfrac{\sqrt{1+v^2}}{av};$$

$\therefore x^{-4}dx(a^2-x^2)^{-\frac{1}{2}} = -\dfrac{1}{a^4}(1+v^2)dv;$

$\therefore \int x^{-4}dx(a^2-x^2)^{-\frac{1}{2}} = -\dfrac{1}{a^4}(v+\tfrac{1}{3}v^3).$

And substituting for v its value $\dfrac{\sqrt{a^2-x^2}}{x}$, we have

$$u = \int x^{-4}dx(a^2-x^2)^{-\frac{1}{2}} = -\dfrac{(a^2+2x^2)(a^2-x^2)^{\frac{1}{2}}}{3a^4x^3} + C.$$

Ex. 77. Integrate $du = \dfrac{x^3 dx}{\sqrt{a^2-x^2}}$.

Put $\sqrt{a^2-x^2} = v$;

$\therefore a^2 - x^2 = v^2$, or $x^2 = a^2 - v^2$;

$\therefore x^4 = a^4 - 2a^2v^2 + v^4$,

$x^3 dx = -a^2 v\, dv + v^3 dv$;

$\therefore \dfrac{x^3 dx}{\sqrt{a^2-x^2}} = -a^2 dv + v^2 dv.$

Integrating and putting for v its value $\sqrt{a^2-x^2}$, we shall have

$$u = \int \dfrac{x^3 dx}{\sqrt{a^2-x^2}} = -\dfrac{(2a^2+x^2)(a^2-x^2)^{\frac{1}{2}}}{3} + C.$$

Ex. 78. Integrate $du = \dfrac{x^5 dx}{a^2+x^2} = x^5 dx(a^2+x^2)^{-1}$.

Put $a^2 + x^2 = v^2$;

$\therefore x^2 = v^2 - a^2$,

and $x^6 = v^6 - 3v^4 a^2 + 3v^2 a^4 - a^6$;

$\therefore x^5 dx = (v^5 - 2a^2 v^3 + a^4 v) dv$;

$\therefore \int \dfrac{x^5 dx}{a^2+x^2} = \int v^3 dv - 2a^2 \int v\, dv + a^4 \int \dfrac{dv}{v}.$

Integrating and putting $\sqrt{a^2+x^2}$ for v, we have

$$u = \int \dfrac{x^5 dx}{a^2+x^2} = \dfrac{(a^2+x^2)(x^2-3a^2)}{4} + a^4 \log.\sqrt{a^2+x^2} + C.$$

INTEGRATION OF RATIONAL AND IRRATIONAL FRACTIONS.

194. Every rational fraction may be resolved into partial fractions by the method of Indeterminate Coefficients (Algebra, Art. 144).

Ex. 79. Integrate $du = \dfrac{a\, dx}{a^2-x^2}$.

(Algebra, Art. 144), $\dfrac{a}{a^2-x^2} = \dfrac{1}{2(a+x)} + \dfrac{1}{2(a-x)}$;

$\therefore du = \dfrac{dx}{2(a+x)} + \dfrac{dx}{2(a-x)}$;

$\therefore u = \tfrac{1}{2} \int \dfrac{dx}{a+x} + \tfrac{1}{2} \int \dfrac{dx}{a-x},$

INTEGRATION OF RATIONAL FRACTIONS. 257.

$$= \tfrac{1}{2} \log. (a+x) - \tfrac{1}{2} \log. (a-x) + C,$$
$$= \tfrac{1}{2} \log. \frac{a+x}{a-x} + C,$$
$$= \log. \sqrt{\frac{a+x}{a-x}} + C.$$

Ex. 80. Integrate $\dfrac{(3x-5)dx}{x^2-13x+40}.$

Ans. $\tfrac{19}{3}\log. (x-8) - \tfrac{10}{3}\log. (x-5) + C.$

Ex. 81. Integrate $\dfrac{(2x+3)dx}{x^3+x^2-2x}.$

Ans. $\tfrac{5}{3}\log. (x-1) - \tfrac{1}{6}\log. (x+2) - \tfrac{3}{2}\log. x + C.$

Ex. 82. If $du = \dfrac{(8x^2+2x-6)dx}{x^3-7x+6},$

$u = 6\log. (x-2) - \log. (x-1) + 3\log. (x+3) + C.$

Ex. 83. If $du = \dfrac{(x+2)dx}{x^3-x},$

$u = \tfrac{1}{2}\log. (x+1) + \tfrac{3}{2}\log. (x-1) - 2\log. x + C.$

Ex. 84. If $du = \dfrac{dx}{x^3+x^2+x+1} = \dfrac{dx}{(1+x)(1+x^2)},$

$u = \tfrac{1}{2}\log. (x+1) - \tfrac{1}{4}\log. (x^2+1) + \tfrac{1}{2}\tan.^{-1} x + C.$

Ex. 85. If $du = \dfrac{6dx}{(x-1)^2(x+1)},$

$u = -\tfrac{3}{2}\log. (x-1) - \dfrac{3}{x-1} + \tfrac{3}{2}\log. (x+1) + C.$

Ex. 86. If $du = \dfrac{3xdx}{(x+1)(x^2+1)},$

$u = \tfrac{3}{4}\log. (x^2+1) + \tfrac{3}{2}\tan.^{-1} x - \tfrac{3}{2}\log. (x+1) + C.$

Ex. 87. If $du = \dfrac{(x^2-x+1)dx}{(x+1)(1+x^2)},$

$u = \tfrac{3}{2}\log. (x+1) - \tfrac{1}{4}\log. (1+x^2) - \tfrac{1}{2}\tan.^{-1} x + C.$

Ex. 88. If $du = \dfrac{(a+bx)dx}{(x-1)(x^2+x+1)},$

$u = \tfrac{1}{3}(a+b)\log. \dfrac{x-1}{\sqrt{x^2+x+1}} + \dfrac{b-a}{2}\int \dfrac{dx}{x^2+x+1}.$

Now, to integrate $\tfrac{1}{2}(b-a)\dfrac{dx}{x^2+x+1},$

put $\quad x^2+x+1=(x+\tfrac{1}{2})^2+\tfrac{3}{4}=\tfrac{3}{4}\left(1+\dfrac{(2x+1)^2}{3}\right)$.

Let $\quad \dfrac{(2x+1)^2}{3}=z^2$;

then $\quad \dfrac{2x+1}{\sqrt{3}}=z$, and $dx=\dfrac{dz\sqrt{3}}{2}$;

$$\therefore \tfrac{1}{2}(b-a)\int\dfrac{dx}{\tfrac{3}{4}\left(1+\dfrac{(2x+1)^2}{3}\right)}=\tfrac{1}{2}(b-a)\dfrac{\sqrt{3}}{2}\int\dfrac{dz}{\tfrac{3}{4}(1+z^2)},$$

$$=\tfrac{1}{2}(b-a)\dfrac{2}{\sqrt{3}}\int\dfrac{dz}{1+z^2},$$

$$=\dfrac{b-a}{\sqrt{3}}\tan^{-1}z,$$

$$=\dfrac{b-a}{\sqrt{3}}\tan^{-1}\left(\dfrac{2x+1}{\sqrt{3}}\right)+C.$$

$$\therefore \int\dfrac{(a+bx)dx}{(x-1)(x^2+x+1)}=\tfrac{1}{3}(a+b)\log.\dfrac{x-1}{\sqrt{x^2+x+1}}+$$

$$\dfrac{b-a}{\sqrt{3}}\tan^{-1}\left(\dfrac{2x+1}{\sqrt{3}}\right)+C.$$

195. Integration of Irrational Fractions.

Ex. 89. Integrate $du=\dfrac{dx}{\sqrt{1+x+x^2}}$.

Here $\quad 1+x+x^2=\tfrac{3}{4}+(\tfrac{1}{2}+x)^2$;

$$\therefore u=\int\dfrac{dx}{\sqrt{1+x+x^2}}=\int\dfrac{dx}{\sqrt{\tfrac{3}{4}+(\tfrac{1}{2}+x)^2}}.$$

Put $\quad \tfrac{1}{2}+x=v$;
then $\quad dx=dv,$
and $\quad u=\int\dfrac{dv}{\sqrt{\tfrac{3}{4}+v^2}}=\log.\{v+\sqrt{\tfrac{3}{4}+v^2}\},$

$$=\log.(\tfrac{1}{2}+x+\sqrt{1+x+x^2})+C.$$

Ex. 90. Integrate $du=\dfrac{dx}{\sqrt{1+x-x^2}}$.

Here $\quad 1+x-x^2=\tfrac{5}{4}-(x-\tfrac{1}{2})^2$;

INTEGRATION OF IRRATIONAL FRACTIONS. 259

$$\therefore u = \int \frac{dx}{\sqrt{1+x-x^2}} = \int \frac{dx}{\sqrt{\frac{5}{4}-(x-\frac{1}{2})^2}}.$$

Put $\qquad x - \frac{1}{2} = v$;
then $\qquad dv = dx$,
and $\qquad u = \int \frac{dx}{\sqrt{1+x-x^2}} = \int \frac{dv}{\sqrt{\frac{5}{4}-v^2}} = \sin.^{-1} \frac{v}{\frac{\sqrt{5}}{2}},$

$$= \sin.^{-1} \frac{2x-1}{\sqrt{5}} + C.$$

Ex. 91. Integrate $du = \dfrac{dx}{\sqrt{1-x-x^2}}$.

Let $\qquad x + \frac{1}{2} = v$;
then $\qquad dx = dv$,
and $\qquad x^2 + x + \frac{1}{4} = v^2$, or $v^2 - \frac{1}{4} = x + x^2$,
or, by changing the signs,
$$\tfrac{1}{4} - v^2 = -x - x^2.$$
If we now add unity to both members, we shall have
$$\tfrac{5}{4} - v^2 = 1 - x - x^2;$$
$$\therefore u = \int \frac{dv}{\sqrt{\frac{5}{4}-v^2}} = \sin.^{-1} \frac{2v}{\sqrt{5}} = \sin.^{-1} \frac{2x+1}{\sqrt{5}} + C.$$

Ex. 92. Integrate
$$du = \frac{dx}{\sqrt{a+bx+cx^2}} = \frac{dx}{\sqrt{\frac{a}{c}+\frac{bx}{c}+x^2}} \frac{1}{\sqrt{c}}.$$

If we put $\dfrac{a}{c} = m$, and $\dfrac{b}{c} = n$, it reduces to

$$du = \frac{1}{\sqrt{c}} \frac{dx}{\sqrt{m+nx+x^2}}.$$

Let $\qquad x + \dfrac{n}{2} = v$;
then $\qquad dx = dv$,
and $\qquad x^2 + nx + \dfrac{n^2}{4} = v^2$,
or $\qquad x^2 + nx + \dfrac{n^2}{4} + m = v^2 + m$,
or $\qquad x^2 + nx + m = v^2 + m - \dfrac{n^2}{4}.$

Hence
$$u = \frac{1}{\sqrt{c}} \int \frac{dv}{\sqrt{\left(m - \frac{n^2}{4}\right) + v^2}} = \frac{1}{\sqrt{c}} \log \left(v + \sqrt{m - \frac{n^2}{4} + v^2}\right),$$
$$= \frac{1}{\sqrt{c}} \log \left(x + \frac{b}{2c} + \sqrt{x^2 + \frac{b}{c}x + \frac{a}{c}}\right) + C.$$

Ex. 93. Integrate $du = \dfrac{dx}{x\sqrt{a + bx + cx^2}}$.

Let
$$v = \frac{1}{x};$$
then
$$-\frac{dv}{v} = \frac{dx}{x};$$
$$\therefore du = \frac{-dv}{\sqrt{av^2 + bv + c}} = \frac{1}{\sqrt{a}} \frac{dv}{\sqrt{v^2 + \frac{b}{a}v + \frac{c}{a}}},$$
$$u = \frac{1}{\sqrt{a}} \int \frac{dv}{\sqrt{v^2 + nv + m}}.$$

The same form as Ex. 92.

INTEGRATION BY PARTS.

196. A most important process for integrating is that which is called *Integration by Parts*, and it depends on the following consideration:

Let u and v be functions of x; then
$$d.(uv) = udv + vdu;$$
$$\therefore uv = \int udv + \int vdu,$$
or
$$\int udv = uv - \int vdu.$$

Hence it appears that, *Having resolved a differential into two factors, one of which can be immediately integrated, we may take this integral regarding the other factor as constant. We must then differentiate the result thus obtained upon the supposition that the factor which we considered constant is the only variable, and then subtract the integral of this differential from the first result.*

Let us take the binomial differential
$$x^{m-1}dx(a + bx^n)^p.$$

If we multiply this by the two factors x^n and x^{-n}, we

INTEGRATION BY PARTS. 261

shall not alter its value, but we shall bring it to the form
$$x^{m-n}x^{n-1}dx(a+bx^n)^p,$$
in which the factor $x^{n-1}dx(a+bx^n)^p$ can be immediately integrated by Art. 186. Put this factor equal to dv; then
$$x^{n-1}dx(a+bx^n)^p = dv;$$
$$\therefore v = \frac{(a+bx^n)^{p+1}}{bn(p+1)}.$$

Again: let $x^{m-n} = u;$
then $(m-n)x^{m-n-1}dx = du;$
$$\therefore \int x^{m-1}dx(a+bx^n)^p = $$
$$\frac{x^{m-n}(a+bx^n)^{p+1}}{bn(p+1)} - \frac{m-n}{bn(p+1)}\int x^{m-n-1}dx(a+bx^n)^{p+1}.$$

Taking the factor $\int x^{m-n-1}dx(a+bx^n)^{p+1}$ and decomposing it, we have
$$\int x^{m-n-1}dx(a+bx^n)^p(a+bx^n),$$
and multiplying, it becomes
$$a\int x^{m-n-1}dx(a+bx^n)^p + b\int x^{m-1}dx(a+bx^n)^p.$$
Substituting these in the preceding equation, we have
$$\int x^{m-1}dx(a+bx^n)^p = $$
$$\frac{x^{m-n}(a+bx^n)^{p+1}}{bn(p+1)} - \frac{a(m-n)}{bn(p+1)}\int x^{m-n-1}dx(a+bx^n)^p,$$
$$- \frac{b(m-n)}{bn(p+1)}\int x^{m-1}dx(a+bx^n)^p.$$

Transposing the last term into the first member, we have
$$\left[1+\frac{b(m-n)}{bn(p+1)}\right]\int x^{m-1}dx(a+bx^n)^p = $$
$$\frac{x^{m-n}(a+bx^n)^{p+1}}{bn(p+1)} - \frac{a(m-n)}{bn(p+1)}\int x^{m-n-1}dx(a+bx^n)^p.$$

Reducing, we obtain *Formula A*,
$$\int x^{m-1}dx(a+bx^n)^p = $$
$$\frac{x^{m-n}(a+bx^n)^{p+1}}{b(pn+m)} - \frac{a(m-n)}{b(pn+m)}\int x^{m-n-1}dx(a+bx^n)^p.$$

Ex. 94. If $du = \dfrac{x^2 dx}{\sqrt{R^2-x^2}} = x^2 dx(R^2-x^2)^{-\frac{1}{2}}.$

Applying Formula A, we have
$m-1=2;\ \therefore m=3;\ a=R^2;\ b=-1;\ n=2;\ p=-\frac{1}{2};$
$$\int x^2 dx(R^2-x^2)^{-\frac{1}{2}} = -\frac{x(R^2-x^2)^{\frac{1}{2}}}{2} + \frac{R^2}{2}\int \frac{dx}{\sqrt{R^2-x^2}}.$$

But by Art. 189, Ex. 49,
$$\int \frac{dx}{\sqrt{R^2-x^2}} = \sin^{-1}\frac{x}{R} + C;$$
$$\therefore \int \frac{x^2 dx}{\sqrt{R^2-x^2}} = -\tfrac{1}{2}x\sqrt{R^2-x^2} + \tfrac{1}{2}R^2 \sin^{-1}\left(\frac{x}{R}\right) + C.$$

Ex. 95. If $du = \dfrac{x^4 dx}{\sqrt{R^2-x^2}} = x^4 dx(R^2-x^2)^{-\frac{1}{2}}$,
$$u = -\tfrac{1}{4}x^3\sqrt{R^2-x^2} + \tfrac{3}{4}R^2 \int \frac{x^2 dx}{\sqrt{R^2-x^2}}.$$

We should then apply Formula A to the last term, and thus find the whole integral.

We see that each process diminishes the exponent of the variable without the parenthesis by the exponent of the variable within. *Formula A is therefore used to diminish the exponent of the variable outside of the parenthesis when that exponent is positive.*

197. We shall now obtain a formula for diminishing the exponent of the parenthesis.

Taking the binomial differential
$$x^{m-1} dx (a+bx^n)^p,$$
we can put this in the form
$$x^{m-1} dx (a+bx^n)^{p-1}(a+bx^n),$$
and by actual multiplication we have
$$\int x^{m-1} dx (a+bx^n)^p =$$
$$a\int x^{m-1} dx (a+bx^n)^{p-1} + b\int x^{m+n-1} dx (a+bx^n)^{p-1}.$$

Now, integrating the second term on the right by Formula A, we shall have
$$m-1 = m+n-1 \text{ or } m = m+n, \text{ and } p = p-1;$$
$$\therefore \int x^{m+n-1} dx (a+bx^n)^{p-1} =$$
$$\frac{x^m(a+bx^n)^p}{b(pn+m)} - \frac{am}{b(pn+m)}\int x^{m-1} dx (a+bx^n)^{p-1}.$$

Substituting this in the preceding equation and reducing, we have *Formula B*,
$$\int x^{m-1} dx (a+bx^n)^p =$$
$$\frac{x^m(a+bx^n)^p}{pn+m} + \frac{pna}{pn+m}\int x^{m-1} dx (a+bx^n)^{p-1}.$$

Ex. 96. Integrate $du = dx\sqrt{r^2-x^2}$.

By applying Formula B, we shall have
$$m-1 = 0; \therefore m = 1; a = r^2; n = 2; p = \tfrac{1}{2}.$$

Then $\int dx(r^2-x^2)^{\frac{1}{2}} = \frac{x}{2}(r^2-x^2)^{\frac{1}{2}} + \frac{r^2}{2}\int dx(r^2-x^2)^{-\frac{1}{2}}.$

But (Art. 189),

$$\int dx(r^2-x^2)^{-\frac{1}{2}} = \int \frac{dx}{\sqrt{r^2-x^2}} = \sin.^{-1}\left(\frac{x}{r}\right);$$

$$\therefore \int dx(r^2-x^2)^{\frac{1}{2}} = \frac{x}{2}(r^2-x^2)^{\frac{1}{2}} + \frac{r^2}{2}\sin.^{-1}\left(\frac{x}{r}\right) + C.$$

Ex. 97. Integrate $du = dx(a^2+x^2)^{\frac{1}{2}}$.

Ans. $u = \frac{x}{2}(a^2+x^2)^{\frac{1}{2}} + \frac{a^2}{2}\int dx(a^2+x^2)^{-\frac{1}{2}}.$

But (Ex. 64),

$$\int dx(a^2+x^2)^{-\frac{1}{2}} = \int \frac{dx}{\sqrt{a^2+x^2}} = \log.(x+\sqrt{a^2+x^2});$$

$$\therefore u = \frac{x}{2}(a^2+x^2)^{\frac{1}{2}} + \frac{a^2}{2}\log.(x+\sqrt{a^2+x^2}) + C.$$

198. We will now obtain two formulas: one for diminishing the exponent of the variable outside the parenthesis when that exponent is negative, the other for diminishing the exponent of the parenthesis when that exponent is negative also.

Take Formula A,
$$\int x^{m-1}dx(a+bx^n)^p =$$
$$\frac{x^{m-n}(a+bx^n)^{p+1}}{b(pn+m)} - \frac{a(m-n)}{b(pn+m)}\int x^{m-n-1}dx(a+bx^n)^p.$$

Reducing, we have
$$\int x^{m-n-1}dx(a+bx^n)^p =$$
$$\frac{x^{m-n}(a+bx^n)^{p+1}}{a(m-n)} - \frac{b(pn+m)}{a(m-n)}\int x^{m-1}dx(a+bx^n)^p;$$

or, substituting $-m+n$ for m, we have *Formula C*,
$$\int x^{-m-1}dx(a+bx^n)^p =$$
$$-\frac{x^{-m}(a+bx^n)^{p+1}}{am} - \frac{b(m-n-np)}{am}\int x^{-m+n-1}dx(a+bx^n)^p,$$

in which the negative sign has been attributed to the exponent m.

Ex. 98. If $du = x^{-5}dx(1+x^2)^{-\frac{1}{2}},$
then $-m-1 = -5;$
$\therefore -m = -4$, and $n=2$, $a=1$, $b=1$, $p=-\frac{1}{2}$;

$$\therefore \int x^{-5}dx(1+x^2)^{-\frac{1}{2}} = -\frac{x^{-4}(1+x^2)^{\frac{1}{2}}}{4} - \tfrac{3}{4}\int x^{-3}dx(1+x^2)^{-\frac{1}{2}}.$$

Apply Formula C again to $\int x^{-3}dx(1+x^2)^{-\frac{1}{2}}$.
Here $-m-1=-3$; therefore $-m=-2$, the other letters being the same as before,

$$\tfrac{3}{4}\int x^{-3}dx(1+x^2)^{-\frac{1}{2}} =$$

$$\tfrac{3}{4}\left[-\frac{x^{-2}(1+x^2)^{\frac{1}{2}}}{2} - \tfrac{1}{2}\int x^{-1}dx(1+x^2)^{-\frac{1}{2}}\right].$$

We have $\int x^{-1}dx(1+x^2)^{-\frac{1}{2}}$, or, in another form,
$$\int \frac{dx}{x\sqrt{1+x^2}}, \text{ to integrate.}$$

Put $\qquad x = \frac{1}{v}; \quad \therefore dx = -\frac{dv}{v^2}.$

Substituting these, we have

$$\int \frac{dx}{x\sqrt{1+x^2}} = -\int \frac{dv}{\sqrt{1+v^2}} = -\log.\left(\frac{1+\sqrt{1+x^2}}{x}\right);$$

$$\therefore \int x^{-5}dx(1+x^2)^{-\frac{1}{2}} =$$
$$-(1+x^2)^{\frac{1}{2}}\left[\frac{1}{4x^4} - \frac{3\cdot 1}{4\cdot 2x^2}\right] - \frac{3\cdot 1}{4\cdot 2}\log.\left(\frac{1+\sqrt{1+x^2}}{x}\right) + C.$$

199. To find a formula for diminishing the exponent of the parenthesis when that exponent is negative.

Take Formula B,
$$\int x^{m-1}dx(a+bx^n)^p =$$
$$\frac{x^m(a+bx^n)^p}{pn+m} + \frac{anp}{pn+m}\int x^{m-1}dx(a+bx^n)^{p-1};$$
$$\therefore (pn+m)\int x^{m-1}dx(a+bx^n)^p =$$
$$x^m(a+bx^n)^p + anp\int x^{m-1}dx(a+bx^n)^{p-1};$$
$$\therefore \int x^{m-1}dx(a+bx^n)^{p-1} =$$
$$-\frac{x^m(a+bx^n)^p}{anp} + \frac{pn+m}{anp}\int x^{m-1}dx(a+bx^n)^p.$$

And substituting $-p+1$ for p, we have *Formula D*,
$$\int x^{m-1}dx(a+bx^n)^{-p} =$$
$$\frac{x^m(a+bx^n)^{-p+1}}{na(p-1)} - \frac{m+n-np}{na(p-1)}\int x^{m-1}dx(a+bx^n)^{-p+1}.$$

When $p=1$, this formula is inapplicable.

200. Let it be required to obtain a formula for integrating
$$\frac{x^m dx}{\sqrt{2ax-x^2}} = x^m dx (2ax-x^2)^{-\frac{1}{2}}.$$
This may be put under the form
$$\int x^{m-\frac{1}{2}} dx (2a-x)^{-\frac{1}{2}}.$$
Applying Formula A to this, we have
$$m-1 = m-\tfrac{1}{2};$$
$$\therefore m = m+\tfrac{1}{2},\ a = 2a,\ b = -1,\ n = 1,\ p = -\tfrac{1}{2};$$
$$\therefore \int x^{m-\frac{1}{2}} dx (2a-x)^{-\frac{1}{2}} =$$
$$-\frac{x^{m-\frac{1}{2}}(2a-x)^{\frac{1}{2}}}{m} + \frac{2a(m-\tfrac{1}{2})}{m} \int x^{m-\frac{3}{2}} dx (2a-x)^{-\frac{1}{2}}.$$
Now, passing the fractional powers of x within the parenthesis, we obtain *Formula E*,
$$\int \frac{x^m dx}{\sqrt{2ax-x^2}} = -\frac{x^{m-1}(2ax-x^2)^{\frac{1}{2}}}{m} + \frac{(2m-1)a}{m} \int \frac{x^{m-1} dx}{\sqrt{2ax-x^2}}.$$
We shall have, if m is a whole number and positive, after m integrations,
$$\int \frac{dx}{\sqrt{2ax-x^2}} = \text{versin.}^{-1} \frac{x}{a} \quad (\text{Art. 189}).$$

INTEGRATION OF LOGARITHMIC FUNCTIONS.

201. We can integrate but few of these forms by any general process. An approximation, however, may always be had by the method of series; but we should never resort to the method of series until all our efforts to obtain an exact integral fail.

Let it be required to integrate
$$f(x) dx\ (\log. x)^n.$$
If, in the formula
$$\int u\, dv = uv - \int v\, du, \qquad (1)$$
we make
$u = (\log. x)^n$, and $dv = f(x) dx$, or $v = \int f(x) dx$,
and substitute these in (1), we shall have *Formula F*,

$$\int f(x)dx \,(\log. x)^n = (\log. x)^n \int f(x)dx - \int f(x)dx \int d \cdot (\log. x)^n. \quad (2)$$

If n is a positive whole number, the successive application of this formula will finally reduce the integration of the proposed form to that of an algebraic function; then the proposed expression will be integrable, provided we can integrate, in succession, the algebraic functions which enter into it.

Ex. 99. Let us take, as an example,

$$du = \frac{x \log. x \, dx}{\sqrt{a^2 + x^2}}. \quad (1)$$

In this expression, we have $n = 1$.

Put $\qquad u = \log. x,$

and $\qquad dv = \dfrac{x\, dx}{\sqrt{a^2+x^2}};$

then $\qquad du = \dfrac{dx}{x},$

and $\qquad v = \displaystyle\int \dfrac{x\, dx}{\sqrt{a^2+x^2}} = \sqrt{a^2+x^2}$ (Art. 187).

Substituting these in Formula F, we have

$$\int \frac{x \log. x \, dx}{\sqrt{a^2+x^2}} = \sqrt{a^2+x^2} \cdot \log. x - \int \frac{\sqrt{a^2+x^2}}{x} \cdot dx.$$

To integrate the second term,

$$-\int \frac{\sqrt{a^2+x^2}}{x} \cdot dx,$$

we shall multiply it by $\dfrac{\sqrt{a^2+x^2}}{\sqrt{a^2+x^2}}$,

which will reduce it to $\displaystyle\int \dfrac{(a^2+x^2)dx}{x\sqrt{a^2+x^2}},$

or $\qquad a^2 \displaystyle\int \dfrac{dx}{x\sqrt{a^2+x^2}} + \int \dfrac{x\, dx}{\sqrt{a^2+x^2}}.$

Now $\qquad \displaystyle\int \dfrac{x\, dx}{\sqrt{a^2+x^2}} = \sqrt{a^2+x^2}$ (Art. 186).

The other can be integrated in a manner entirely similar to that of Ex. 98 by assuming $x = \dfrac{a}{v}$; but we shall do it differently. Thus:

INTEGRATION OF LOGARITHMIC FUNCTIONS. 267

In the expression $\dfrac{dx}{x\sqrt{a^2+x^2}}$,

put $\sqrt{a^2+x^2}=v$;

$\therefore a^2+x^2=v^2$, and $x=\sqrt{v^2-a^2}$,

and $xdx=vdv$; $\therefore dx=\dfrac{vdv}{\sqrt{v^2-a^2}}$,

or $\dfrac{dx}{x\sqrt{a^2+x^2}}=\dfrac{dv}{v^2-a^2}$,

by dividing the second member by $v\sqrt{v^2-a^2}=vx$.

$$\therefore \int\dfrac{dx}{x\sqrt{a^2+x^2}}=\int\dfrac{dv}{v^2-a^2}=-\dfrac{1}{a}\cdot\tfrac{1}{2}\log.\left(\dfrac{v+a}{x}\right)^2$$

(Art. 194).

$$=-\dfrac{1}{a}\cdot\log.\dfrac{v+a}{x},$$

$$=-\dfrac{1}{a}\cdot\log.\dfrac{\sqrt{a^2+x^2}+a}{x};$$

$$\therefore a^2\int\dfrac{dx}{x\sqrt{a^2+x^2}}=-a\cdot\log.\dfrac{\sqrt{a^2+x^2}+a}{x}.$$

Substituting these in (1), we have

$$\int\dfrac{x\log. xdx}{\sqrt{a^2+x^2}}=\sqrt{a^2+x^2}\log. x+$$

$$a\log.\dfrac{\sqrt{a^2+x^2}+a}{x}-\sqrt{a^2+x^2}+C.$$

202. A very useful case of the above formula is that in which $f(x)=x^m$, the form being

$$\int x^m (\log. x)^n dx.$$

In the expression

$$\int udv=uv-\int vdu, \qquad (1)$$

let $u=(\log. x)^n$,

and $dv=x^m dx$;

then $du=n(\log. x)^{n-1}\dfrac{dx}{x}$, and $v=\dfrac{x^{m+1}}{m+1}$.

Substitute these in (1), then

$$\int x^m (\log. x)^n dx=$$

$$\dfrac{x^{m+1}}{m+1}(\log. x)^n-\dfrac{n}{m+1}\int x^m (\log. x)^{n-1} dx.$$

Again: $\int x^m (\log. x)^{n-1} dx =$
$$\frac{x^{m+1}}{m+1}(\log. x)^{n-1} - \frac{n-1}{m+1}\int x^m (\log. x)^{n-2} dx,$$
and $\int x^m (\log. x)^{n-2} dx =$
$$\frac{x^{m+1}}{m+1}(\log. x)^{n-2} - \frac{n-2}{m+1}\int x^m (\log. x)^{n-3} dx.$$

The law of the series becomes manifest. Hence, substituting for the integrals on the right their values as obtained in the successive equations, we have *Formula G*,
$$\int x^m (\log. x)^n dx =$$
$$x^{m+1}\left[\frac{(\log. x)^n}{m+1} - \frac{n}{(m+1)^2}(\log. x)^{n-1} + \frac{n(n-1)}{(m+1)^3}(\log. x)^{n-2} - \frac{n(n-1)(n-2)}{(m+1)^4}(\log. x)^{n-3} +, \text{etc.}\right] + C.$$

When n is a positive whole number, this formula will reduce the exponent of the logarithm; but if m be equal to *minus unity*, its application fails. In this case we should have
$$\frac{(\log. x)^n dx}{x} = (\log. x)^n . d . \log. x;$$
$$\therefore \int (\log. x)^n . \frac{dx}{x} = \frac{(\log. x)^{n+1}}{n+1} + C.$$

203. If n be negative, we have
$$\int \frac{dx}{x(\log. x)^n} = -\frac{(\log. x)^{-(n-1)}}{n-1},$$
so that the formula for integrating by parts gives *Formula H*,
$$\int \frac{x^m dx}{(\log. x)^n} = -\frac{x^{m+1}}{n-1}\left[\frac{1}{(\log. x)^{n-1}} - \frac{m+1}{n-2} \cdot \frac{1}{(\log. x)^{n-2}} + \frac{(m+1)^2}{(n-2)(n-3)} \cdot \frac{1}{(\log. x)^{n-3}} +, \text{etc.}\right] +$$
$$\frac{(m+1)^{n-1}}{1.2.3\ldots(n-1)}\int \frac{x^m dx}{\log. x}.$$

This formula ceases to be applicable when n becomes equal to unity.

Ex. 100. Let $du = x^3 (\log. x)^2 dx$.
Here $m = 3, n = 2$; and, applying Formula G,
$$u = \frac{x^4}{4}\{(\log. x)^2 - \tfrac{1}{2}(\log. x) + \tfrac{1}{8}\} + C.$$

INTEGRATION OF EXPONENTIAL FUNCTIONS. 269

Ex. 101. Let $du = \dfrac{x^4 dx}{(\log. x)^2}.$

Applying Formula H, we have $m=4, n=2.$

$$\therefore u = -\dfrac{x^5}{\log. x} + 5\int \dfrac{x^4 dx}{\log. x}.$$

This last integral can not be obtained except by series.
Thus:
Put $\qquad z = x^5;$

$$\therefore x^4 dx = \dfrac{dz}{5};$$

also $\qquad \log. z = 5 \log. x;$

$$\therefore \dfrac{x^4 dx}{\log. x} = \dfrac{dz}{\log. z}.$$

Now, let $\qquad \log. z = y;$

$\therefore z = e^y,$ and $dz = e^y . dy.$

But, by Art. 70, $z = e^y = 1 + y + \tfrac{1}{2}y^2 + \tfrac{1}{6}y^3 +,$ etc.;

$$\therefore \int \dfrac{dz}{\log. z} =$$

$\int \dfrac{e^y dy}{y} = \int \dfrac{dy}{y} + \int dy + \tfrac{1}{2}\int y dy + \tfrac{1}{6}\int y^2 dy +,$ etc.,

$= \log. y + y + \tfrac{1}{4}y^2 + \tfrac{1}{18}y^3 +,$ etc.,

$= \log.^2 z + \log. z + \tfrac{1}{4}(\log. z)^2 + \tfrac{1}{18}(\log. z)^3 +,$ etc.

Substitute for z its value x^5, and we shall have the required integral.

204. INTEGRATION OF EXPONENTIAL FUNCTIONS.

Let it be required to integrate
$\qquad x^m . a^x dx.$

Put $\qquad u = x^m;$
then $\qquad du = mx^{m-1} dx;$
put $\qquad v = \int a^x dx;$
then $\qquad v = \dfrac{a^x}{\log. a}.$

Substitute these in the formula
$$\int u dv = uv - \int v du,$$

and we have $\int x^m a^x dx = \dfrac{x^m . a^x}{\log. a} - \dfrac{m}{\log. a}\int x^{m-1} a^x dx.$

Again: $\int x^{m-1} a^x dx = \dfrac{x^{m-1} . a^x}{\log. a} - \dfrac{m-1}{\log. a}\int x^{m-2} a^x dx,$

and $\quad \int x^{m-2} a^x dx = \dfrac{x^{m-2} \cdot a^x}{\log. a} - \dfrac{m-2}{\log. a}\int x^{m-3} a^x dx,$

etc. = etc.

Therefore, by substitution, we have *Formula I*,

$$\int x^m a^x dx = \dfrac{a^x}{\log. a}\left[x^m - \dfrac{mx^{m-1}}{\log. a} + \dfrac{m(m-1)x^{m-2}}{(\log. a)^2} - \text{etc.} \pm \dfrac{1.2.3\ldots m}{(\log. a)^m}\right] + C.$$

The upper sign of the last term having place when m is even, and the lower when m is odd.

205. The series within the brackets does not terminate when m is *negative*, and is therefore inapplicable. But that case, if the expression is $x^{-m} a^x dx$,

put $\quad u = a^x,$
and $\quad dv = x^{-m} dx;$
$\therefore du = a^x l a dx,$
and $\quad v = -\dfrac{x^{-m+1}}{m-1} = -\dfrac{1}{(m-1)x^{m-1}}.$

Substituting these in the formula

$$\int u dv = uv - \int v du,$$

we have $\quad \displaystyle\int \dfrac{a^x dx}{x^m} = -\dfrac{a^x}{(m-1)x^{m-1}} + \dfrac{\log. a}{m-1}\int \dfrac{a^x dx}{x^{m-1}}.$

Again: $\quad \displaystyle\int \dfrac{a^x dx}{x^{m-1}} = -\dfrac{a^x}{(m-2)x^{m-2}} + \dfrac{\log. a}{m-2}\int \dfrac{a^x dx}{x^{m-2}},$

and $\quad \displaystyle\int \dfrac{a^x dx}{x^{m-2}} = -\dfrac{a^x}{(m-3)x^{m-3}} + \dfrac{\log. a}{m-2}\int \dfrac{a^x dx}{x^{m-3}}.$

Therefore, by substitution, we have *Formula K*,

$$\int \dfrac{a^x dx}{x^m} = -\dfrac{a^x}{(m-1)x^{m-1}}\left[1 + \dfrac{\log. a}{m-2}x + \dfrac{(\log. a)^2}{(m-2)(m-3)}x^2 + \dfrac{(\log. a)^{m-2}}{(m-2)(m-1)\ldots 1}x^{m-2}\right] + \dfrac{(\log. a)^{m-1}}{1.2.3\ldots (m-1)}\int \dfrac{a^x dx}{x}.$$

The integral $\displaystyle\int \dfrac{a^x dx}{x}$ may be obtained by series.

Ex. 102. Integrate $du = x^3 a^x dx.$

Using Formula I, we have

$$u = \int x^3 a^x dx = \dfrac{a^x}{\log. a}\left[x^3 - \dfrac{3x^2}{\log. a} + \dfrac{6x}{(\log. a)^2} - \dfrac{6}{(\log. a)^3}\right] + C.$$

Ex. 103. Integrate $du = \dfrac{a^x dx}{x^3}$.

Ans. $\displaystyle\int \dfrac{a^x dx}{x^3} = -\dfrac{a^2}{2x^2}\left[1 + \log. a . x\right] + \dfrac{(\log. a)^2}{2}$
$\{\log. a + \log. a . x + \tfrac{1}{2}(\log. a)^2 +, \text{etc.}\}$.

INTEGRATION OF TRIGONOMETRICAL FUNCTIONS.

206. All functions of trigonometrical lines may be reduced to functions of the sine and cosine; we shall therefore confine our attention principally to these.

Ex. 104. The expression
$$\sin.^m x\, dx,$$
when m is an entire quantity, may be integrated thus:
Put $\quad\quad\quad\quad \sin. x = v\,;$
then $\quad\quad\quad\quad x = \sin.^{-1} v,$
and $\quad\quad\quad\quad dx = \dfrac{dv}{\sqrt{1-v^2}}\,;$

$\therefore \sin.^m x\, dx = \dfrac{v^m dv}{\sqrt{1-v^2}},$

which can be integrated by the preceding formulas A, B, C, or D.

Ex. 105. Integrate $\quad \sin.^m x \cos.^n x\, dx$.
Put $\quad\quad\quad\quad \sin. x = v\,;$
then $\quad\quad\quad\quad \cos. x = \sqrt{1-v^2}\,;$

$\therefore dx = \dfrac{dv}{\cos. x} = \dfrac{dv}{\sqrt{1-v^2}},$

and $\quad \int \sin.^m x \cos.^n x\, dx = \int v^m dv (1-v^2)^{\frac{n-1}{2}}$,
which can be integrated by the formulas A, B, C, or D.

Ex. 106. Integrate $\dfrac{dx}{\sin. x}.$
Put $\quad\quad\quad\quad \cos. x = v\,;$
then $\quad\quad\quad\quad dx = -\displaystyle\int \dfrac{dv}{1-v^2}.$

This, integrated by rational fractions, gives

$\displaystyle\int \dfrac{dx}{\sin. x} = \log. \left(\dfrac{1-\cos. x}{1+\cos. x}\right)^{\frac{1}{2}} = \log. \tan. \tfrac{1}{2}x.$

(Trig., Prob. XV.)

Ex. 107. Integrate $\sin^3 x\, dx$.
Put $\cos x = v$;
then $\sin x = \sqrt{1-v^2}$,
and $-\sin x\, dx = dv$.
$$dx = -\frac{dv}{\sin x} = -\frac{dv}{\sqrt{1-v^2}};$$
$$\therefore \int \sin^3 x\, dx = -\int dv(1-v^2),$$
$$= -v + \tfrac{1}{3}v^3,$$
$$= -\cos x + \tfrac{1}{3}\cos^3 x,$$
$$= -\cos x + \tfrac{1}{3}\cos x(1-\sin^2 x),$$
$$= -\tfrac{2}{3}\cos x - \tfrac{1}{3}\cos x \sin^2 x + C.$$

Ex. 108. Integrate $\cos^3 x\, dx$.
$$\int \cos^3 x\, dx = \int \cos x\, dx (1 - \sin^2 x),$$
$$= \sin x - \int \cos x \sin^2 x\, dx,$$
$$= \tfrac{2}{3}\sin x + \tfrac{1}{3}\sin x \cos^2 x + C.$$

Ex. 109. Integrate $\dfrac{x \sin^{-1} x\, dx}{(1-x^2)^{\frac{3}{2}}}$.

Put $\sin^{-1} x = u$; $\therefore \dfrac{dx}{\sqrt{1-x^2}} = du$;

and $x\, dx (1-x^2)^{-\frac{3}{2}} = dv$;

$$\therefore v = \frac{1}{(1-x^2)^{\frac{1}{2}}} \text{ (Art. 186)};$$

$$\therefore \int \frac{x \sin^{-1} x\, dx}{(1-x^2)^{\frac{3}{2}}} = \frac{\sin^{-1} x}{(1-x^2)^{\frac{1}{2}}} + \log \sqrt{\frac{1-x}{1+x}} + C.$$

Ex. 110. Integrate $\dfrac{x^2 \tan^{-1} x \cdot dx}{1+x^2}$.

Put $\tan^{-1} x = u$,

and $\dfrac{x^2 dx}{1+x^2} = dv = \left(1 - \dfrac{1}{1+x^2}\right) dx$;

$\therefore \dfrac{dx}{1+x^2} = du$; $v = x - \tan^{-1} x$.

Hence $\displaystyle \int \frac{x^2 \tan^{-1} x\, dx}{1+x^2}$

$= \tan^{-1} x\, (x - \tan^{-1} x) - \displaystyle\int \frac{x\, dx}{1+x^2} + \int \frac{\tan^{-1} x\, dx}{1+x^2}$,

$= \tan^{-1} x (x - \tan^{-1} x) - \tfrac{1}{2} \log (1+x^2) + \tfrac{1}{2}(\tan^{-1} x)^2$,

$= \tan^{-1} x - \tfrac{1}{2}(\tan^{-1} x)^2 - \log \sqrt{1+x^2} + C.$

Integration in Terms of Sines and Cosines of Multiple Arcs.

207. When the exponents of the sine and cosine are positive whole numbers, the integration may be effected without introducing any powers of the trigonometrical lines, the sines and cosines of the multiple arcs occurring instead, and these are more easily calculated than the powers.

In order to do this, we will take the expressions from Plane Trig., Prob. XVI. and XVII., using x for the arc instead of A. Then

$\sin. 2x = 2 \sin. x \cos. x,$ (1)
$\sin. 3x = 3 \sin. x - 4 \sin.^3 x,$ (2)
$\sin. 4x = 8 \cos.^4 x - 4 \sin. x \cos. x,$ (3)
etc. =   etc.,
$\cos. 2x = 2 \cos.^2 x - 1,$ (4)
$\cos. 3x = 4 \cos.^3 x - 3 \cos. x,$ (5)
$\cos. 4x = 8 \cos.^4 x - 8 \cos^2. x + 1,$ (6)
etc. =   etc.

We easily transform these into the following:

$\sin.^2 x = -\frac{1}{2} \cos. 2x + \frac{1}{2},$ (7)
$\sin.^3 x = -\frac{1}{4} \sin. 3x + \frac{3}{4} \sin. x,$ (8)
$\sin.^4 x = \frac{1}{8} \cos. 4x - \frac{1}{2} \cos. 2x + \frac{3}{8},$ (9)
etc. =   etc.,
$\cos.^2 x = \frac{1}{2} \cos. 2x + \frac{1}{2},$ (10)
$\cos.^3 x = \frac{1}{4} \cos. 3x + \frac{3}{4} \cos. x,$ (11)
$\cos.^4 x = \frac{1}{8} \cos. 4x + \frac{1}{2} \cos. 2x + \frac{3}{8}.$ (12)

Multiplying (Eqs. 7-12) by dx and integrating, we have

$\int \sin.^2 x\, dx = -\frac{1}{2} \int \cos. 2x\, dx + \frac{1}{2}x,$
$= -\frac{1}{4} \sin. 2x + \frac{1}{2}x + C.$ (13)
$\int \sin.^3 x\, dx = -\frac{1}{4} \int \sin. 3x\, dx + \frac{3}{4} \int \sin. x\, dx,$
$= \frac{1}{12} \cos. 3x - \frac{3}{4} \cos. x + C.$ (14)
$\int \sin.^4 x\, dx = \frac{1}{8} \int \cos. 4x\, dx - \frac{1}{2} \int \cos. 2x\, dx + \frac{3}{8}x,$
$= \frac{1}{32} \sin. 4x - \frac{1}{4} \sin. 2x + \frac{3}{8}x + C.$ (15)
etc. =   etc.
$\int \cos.^2 x\, dx = \frac{1}{2} \int \cos. 2x\, dx + \frac{1}{2}x,$
$= \frac{1}{4} \sin. 2x + \frac{1}{2}x + C.$ (16)
$\int \cos.^3 x\, dx = \frac{1}{4} \int \cos. 3x\, dx + \frac{3}{4} \int \cos. x\, dx,$
$= \frac{1}{12} \sin. 3x + \frac{3}{4} \sin. x + C.$ (17)

$\int \cos.^4 x\, dx = \frac{1}{8}\int \cos. 4x\, dx + \frac{1}{2}\int \cos. 2x\, dx + \frac{3}{8}x,$
$\qquad = \frac{1}{32}\sin. 4x + \frac{1}{4}\sin. 2x + \frac{3}{8}x + C.$ (18)
etc. = etc.

To integrate the forms

$$X dx \sin.^{-1} x, \qquad (1)$$
$$X dx \cos.^{-1} x, \qquad (2)$$

in which X is an algebraic function of x:

1°. Let $\quad u = \sin.^{-1} x;$

$$\therefore du = \frac{dx}{\sqrt{1-x^2}}.$$

$$dv = X dx; \quad \therefore v = \int X dx.$$

Then $\quad \int u\, dv = uv - \int v\, du$

becomes $\int X dx \sin.^{-1} x = \sin.^{-1} x \int X dx - \int \frac{\int X dx}{\sqrt{1-x^2}} \cdot dx,$

Formula M.

Ex. 111. Integrate $\dfrac{x^3 dx}{\sqrt{1-x^2}} \sin.^{-1} x.$

First integrate $\dfrac{x^3 dx}{\sqrt{1-x^2}}$ by Formula A and Art. 187, and we have

$$\int \frac{x^3 dx}{\sqrt{1-x^2}} = -(\tfrac{1}{3}x^2 + \tfrac{2}{3})\sqrt{1-x^2}.$$

Multiplying this by the $\int \dfrac{dx}{\sqrt{1-x^2}},$ we have

$$-\int (\tfrac{1}{3}x^2 + \tfrac{2}{3}) dx = -\left(\frac{x^3}{9} + \frac{2x}{3}\right).$$

Substituting these in the above formula, we have

$$\int \frac{x^3 dx}{\sqrt{1-x^2}} \sin.^{-1} x =$$
$$-(\tfrac{1}{3}x^2 + \tfrac{2}{3})\sqrt{1-x^2} \sin.^{-1} x + \tfrac{1}{9}x^3 + \tfrac{2}{3}x^2 + C.$$

INTEGRATION BETWEEN LIMITS.

208. In the process of differentiation, we saw that all the constants which were added to or subtracted from the variable quantities disappeared; and, on the contrary, when we performed an integration, we added a constant quantity to correct it, which we represented by the letter C. In order to find the value of this constant, we

must first find what particular value of the variable makes the integral equal to *zero;* we then have two equations each containing C, which quantity may be eliminated by taking the difference between the two equations. Thus, if we have $\int 2xdx = x^2 + C$ for the *general value* of the integral, and the problem indicates that, for the *particular value*, $x=a$, the integral becomes *zero;* then $0 = a^2 + C$. If we subtract this from the general value of the integral, we will have
$$x^2 - a^2,$$
and the constant C has disappeared. This is called the *corrected integral*, and is expressed thus:
$$\int_a^x 2xdx = x^2 - a^2.$$
The value of this integral commences when $x=a$. Now, if we give another value to x, say $x=b$, then we have fully determined the value of the integral, and it is written
$$\int_a^b 2xdx = b^2 - a^2.$$
This is called the *definite integral*, and is said to be taken between the limits $x=b$ and $x=a$; the former is called the superior limit, and the latter the inferior limit, and the operation is called *integration between limits*.

As every function of x may represent the ordinates of a curve whose abscissa is x, it follows that the operation of integrating between limits may be applied to finding the lengths and areas of curves, the surfaces and volumes of solids of revolution, etc.

RECTIFICATION OF PLANE CURVES.

209. The rectification of a curve is the obtaining of a straight line equal to the arc of the curve. When an expression for the length of a curve can be found in a finite number of algebraic terms, the curve is said to be rectifiable.

The differential of the arc of a curve referred to rectangular axes is (Art. 139)
$$dz = \sqrt{dx^2 + dy^2}.$$
Hence, to find the length of a curve given by its equation, we derive the following

Rule.

Differentiate the equation of the curve, and find either the differential of x or the differential of y, and substitute it in the expression

$$dz = \sqrt{dx^2 + dy^2}.$$

Simplify and integrate the expression between the proper limits.

Ex. 112. Find the length of the common parabola
$$y^2 = 2px.$$
Differentiate and divide by 2; we have

$$dx = \frac{y}{p}dy;$$

$$\therefore dx^2 = \frac{1}{p^2}y^2 dy^2.$$

Substituting this in the differential of the arc, we have

$$dz = \sqrt{\frac{y^2 dy^2}{p^2} + dy^2},$$

$$= \frac{1}{p}dy\sqrt{p^2 + y^2};$$

$$\therefore z = \frac{1}{p}\int dy(p^2 + y^2)^{\frac{1}{2}}.$$

Integrating this by Formula B, we shall have

$$z = \tfrac{1}{2}y(p^2 + y^2)^{\frac{1}{2}} + \tfrac{1}{2}p^2 \int \frac{dy}{\sqrt{p^2 + y^2}}.$$

Integrating $\int \dfrac{dy}{\sqrt{p^2 + y^2}}$ like Ex. 64, we have

$$\frac{1}{2p^2}\int \frac{dy}{\sqrt{p^2 + y^2}} = \frac{1}{2p^2}\log.(y + \sqrt{p^2 + y^2}).$$

Therefore, multiplying the integral by the constant factor $\dfrac{1}{p}$, we have

$$z = \frac{y}{2p}\sqrt{p^2 + y^2} + \frac{p}{2}\log.(y + \sqrt{p^2 + y^2}) + C.$$

If we estimate the arc from the vertex of the parabola, we shall have, if $z = 0$, $y = 0$;

$$\therefore 0 = \frac{p}{2}\log. p + C; \quad \therefore C = -\tfrac{1}{2}p\log. p,$$

RECTIFICATION OF PLANE CURVES. 277

and, consequently,
$$z = \frac{y}{2p}\sqrt{p^2+y^2} + \frac{p}{2}\log\left(\frac{y+\sqrt{p^2+y^2}}{p}\right).$$
The value of the arc for a given ordinate can only be found approximately.

Ex. 113. Find the length of the cubical parabola
$$y^2 = p^2 x^3.$$
By differentiating and squaring, we have
$$dy^2 = \frac{9}{4}p^2 x\, dx^2;$$
$$\therefore z = \int dx \left(1 + \frac{9p^2}{4}x\right)^{\frac{1}{2}},$$
$$= \frac{8}{27p^2}\left(1 + \frac{9p^2}{4}x\right)^{\frac{3}{2}} + C.$$
If $x=0$, then $z=0$;
$$\therefore 0 = \frac{8}{27p^2} + C; \quad \therefore C = -\frac{8}{27p^2}.$$
Hence
$$z = \frac{8}{27p^2}\left[\left(1 + \frac{9p^2 x}{4}\right)^{\frac{3}{2}} - 1\right].$$
The cubical parabola is therefore rectifiable.

Ex. 114. Find the length of an arc of the semi-cubical parabola whose equation is
$$y^3 = p^2 x^2.$$
$$\text{Ans. } z = \frac{(9y+4p^2)^{\frac{3}{2}}}{27p} - \frac{8p^2}{27}.$$

Ex. 115. Find the length of the arc of a cycloid.
The differential equation of the cycloid is
$$dx = \frac{y\, dy}{\sqrt{2ry - y^2}};$$
$$\therefore dx^2 = \frac{y\, dy^2}{2r - y}.$$
And $z = \int (dx^2 + dy^2)^{\frac{1}{2}} = \sqrt{2r}\int (2r-y)^{-\frac{1}{2}} dy,$
$= -2\sqrt{2r(2r-y)} + C$, by Art. 184.

Taking this between the limits, $y=0$ and $y=2r$, we have $z = 4r =$ half the arc.

Hence *the length of the cycloid is equal to 8r, or four times the diameter of the generating circle.*

Ex. 116. Determine the length of the curve whose equation is
$$y = \log \frac{e^x+1}{e^x-1},$$
between the limits $x=1$ and $x=2$.

By differentiating and squaring, we have
$$dy^2 = \frac{4e^{2x}dx^2}{(e^{2x}-1)^2};$$
$$\therefore z = \int (dx^2+dy^2)^{\frac{1}{2}} = \int \frac{e^{2x}+1}{e^{2x}-1} dx.$$

Separating this by the method of rational fractions, we have
$$\int \frac{e^{2x}+1}{e^{2x}-1} dx = \int \frac{e^x dx}{e^x+1} + \int \frac{e^x dx}{e^x-1} - \int dx,$$
$$= \log.(e^x+1) + \log.(e^x-1) - x + C$$
$$= \log.(e^{2x}-1) - x + C.$$

Now, when $x=1$, $\quad f = \log.(e^2-1) - 1 + C.$
If $x=2$, $\quad\quad\quad\quad f = \log.(e^4-1) - 2 + C.$
$$\therefore z = \log.(e^2+1) - 1.$$
But $\quad\quad\quad\quad \log. e = 1;$
$$\therefore z = \log.(e^2+1) - \log. e,$$
$$= \log.(e+e^{-1}).$$

Ex. 117. Find the length of the line whose equation is
$$y = ax.$$
$$\textit{Ans. } z = \sqrt{x^2+y^2}.$$

Ex. 118. Find the length of the spiral of Archimedes whose equation is
$$r = a\theta.$$
$$dr = a d\theta; \quad \therefore d\theta^2 = \frac{dr^2}{a^2}.$$
$$dz = \sqrt{dr^2+r^2 d\theta^2},$$
$$= \sqrt{dr^2+r^2 \frac{dr^2}{a^2}},$$
$$= \frac{1}{a} \cdot dr \sqrt{a^2+r^2};$$
$$z = \frac{1}{a} \int dr \sqrt{a^2+r^2},$$
$$= \frac{r\sqrt{a^2+r^2}}{2a} + \frac{a}{2} \log.(r+\sqrt{a^2+r^2}) + C.$$

If $r=0$, $z=0$;

then $0 = \dfrac{a}{2} \log. a + C$;

$$\therefore C = -\dfrac{a}{2} \log. a ;$$

$$\therefore z = \dfrac{r\sqrt{a^2+r^2}}{2a} + \dfrac{a}{2} \log. \dfrac{r+\sqrt{a^2+r^2}}{a}.$$

Ex. 119. Find the length of the curve whose equation is
$$y^{\frac{2}{3}} = R^{\frac{2}{3}} - x^{\frac{2}{3}},$$
from $x=0$ to $x=R$.

Ans. $\frac{3}{2}R$.

The whole length of the four branches is $4 \times \frac{3}{2}R = 6R$.

Ex. 120. To find the length of an arc of the logarithmic spiral.

We have for the equation
$$\theta = \log. r ;$$
$$\therefore d\theta = \dfrac{dr}{r}.$$

Substitute this in $dz = \sqrt{dr^2 + r^2 d\theta^2}$ and reduce, we have $dz = dr\sqrt{2}$;

$$\therefore z = r\sqrt{2} + C.$$

If we estimate the arc from the pole where $r=0$, we shall have
$$z = r\sqrt{2}.$$

Hence *the length of an arc of a Naperian logarithmic spiral, estimated from the pole to any point of the curve, is equal to the diagonal of a square described on the radius vector.*

Ex. 121. *Determine the length of the involute of a circle.*

Let C be the centre of a circle whose radius is r; APR a portion of the *involute*.

Put the angle OCA$=\theta$, and x and y the co-ordinates of the point P, the origin of rectangular axes being at C.

Then OP=arc OA=the portion of the string unwound, and OP$=r\theta$.

Now $\dfrac{OD}{OC} = \sin \theta$; $\therefore OD = r \cdot \sin \theta$, (1)

$\dfrac{CD}{OC} = \cos \theta$; $\therefore CD = r \cdot \cos \theta$. (2)

Also $\dfrac{OB}{OP} = \cos \theta$; $\therefore OB = r\theta \cdot \cos \theta$; (3)

$\dfrac{PB}{OP} = \sin \theta$; $\therefore BP = r\theta \cdot \sin \theta$. (4)

But $x = CE = r \cdot \cos \theta + r\theta \cdot \sin \theta$, (5)
$y = PE = r \cdot \sin \theta - r\theta \cdot \cos \theta$. (6)

If $AP = z$, then $dz = \sqrt{dx^2 + dy^2}$.
Differentiating (5) and (6), we obtain
$dx = -r \sin \theta d\theta + rd\theta \sin \theta + r\theta \cos \theta d\theta$
$ = r\theta \cos \theta d\theta$, (7)
$dy = r \cos \theta d\theta - rd\theta \cos \theta + r\theta \sin \theta d\theta$
$ = r\theta \sin \theta d\theta$. (8)

Substituting these values in the differential of the arc and reducing, we have
$$dz = r\theta d\theta;$$
$$\therefore z = \tfrac{1}{2} r\theta^2.$$

Quadrature of Curves.

210. The *quadrature* of a curve is the expression of its area. When this expression can be found in finite algebraic terms, the curve is quadrable, and may be represented by an equivalent square.

We have shown that the differential of the area of a curve referred to rectangular axes is
$$ydx.$$

Hence, to find the area of a curve when so referred, we have the following

Rule.

From the equation of the curve find the value of y, and multiply it by dx; or, find the differential of x and multiply that by y, then simplify and integrate between the proper limits.

Ex. 122. Let $y = ax$; (1)
then $ydx = axdx$,
$\int ydx = a\int xdx$;
$\therefore$ area $= \tfrac{1}{2}ax^2$.

QUADRATURE OF CURVES. 281

But from (1) we have $xy = ax^2$;
$$\therefore \text{area} = \tfrac{1}{3}xy,$$
which is the area of a triangle—*half the product of the base and perpendicular.*

Ex. 123. Let $y^2 = 2px$;
then $y = \sqrt{2px}$,
$$ydx = (2px)^{\frac{1}{2}}dx;$$
$$\text{area} = (2p)^{\frac{1}{2}} \int x^{\frac{1}{2}} dx,$$
$$= \frac{2(2px)^{\frac{1}{2}}x}{3},$$
$$= \tfrac{2}{3}xy,$$
which is the area of the *common parabola*.

Ex. 124. To find the area of a circle.
We have $y = \sqrt{r^2 - x^2}$;
$$\therefore ydx = dx\sqrt{r^2 - x^2},$$
$$\int ydx = \int dx\sqrt{r^2 - x^2},$$
$$= \tfrac{1}{2}x\sqrt{r^2 - x^2} + \tfrac{1}{2}r^2 \sin^{-1}\frac{x}{r},$$
by Formula B.
$$\therefore \text{area} = \int_0^r dx\sqrt{r^2 - x^2} = \tfrac{1}{4}\pi r^2$$
= area of a quadrant.
Hence area of the circle $= \pi r^2$.

It will be seen by the diagram that the area of any segment of the circle, as ABPC, denoted by
$$\int_0^{x'} y dx,$$
is composed of two parts; the first part being the area of the triangle $\text{ABP} = \tfrac{1}{2}x'\sqrt{r^2 - x'^2}$, the second part the area of the sector $\text{APC} = \tfrac{1}{2}r^2 \sin^{-1}\dfrac{x'}{r}$.

Ex. 125. To find the area of a circle whose radius is unity.

The equation of the circle whose radius is unity is
$$y = \sqrt{1 - x^2} = 1 - \tfrac{1}{2}x^2 - \tfrac{1}{8}x^4 - \tfrac{1}{16}x^6 - \tfrac{5}{128}x^8 -, \text{ etc.};$$
$$\therefore ydx = dx\sqrt{1 - x^2} =$$
$$dx - \tfrac{1}{2}x^2 dx - \tfrac{1}{8}x^4 dx - \tfrac{1}{16}x^6 dx - \tfrac{5}{128}x^8 dx -, \text{ etc.};$$

$$\therefore \text{area} = \int dx \sqrt{1-x^2} =$$
$$x - \tfrac{1}{6}x^3 - \tfrac{1}{40}x^5 - \tfrac{1}{112}x^7 - \tfrac{5}{1152}x^9 -, \text{ etc.} + C.$$

Now, when $x=0$, the area $=0$, and $C=0$.

If the arc be equal to $30°$, $x=\tfrac{1}{2}$, and the area of the segment will be
$$=.5 - .0208333 - .0007812 - .0000698 - .0000085 -, \text{ etc.,}$$
$$=.4783055.$$

But since $x=\tfrac{1}{2}$, $\therefore y=\tfrac{1}{2}\sqrt{3}$,
and the area of the $\triangle = \tfrac{1}{4} \cdot \tfrac{1}{2}\sqrt{3} = .2165063$;
$$\therefore \text{ area of the sector of } 30° = .4783055 - .2165063$$
$$= .2617992,$$
which, multiplied by 12, gives the
$$\text{area of the circle} = 3.14159+.$$

Ex. 126. To find the area of an ellipse.

The equation of the ellipse referred to its centre and axes is
$$y = \frac{B}{A}\sqrt{A^2 - x^2};$$
$$\therefore y\,dx = \frac{B}{A} dx \sqrt{A^2 - x^2},$$
or
$$\text{area} = \frac{B}{A} \int dx \sqrt{A^2 - x^2}.$$

But $\int dx \sqrt{A^2 - x^2} =$ the area of a circle whose radius is A. That is, it is the area of a circle described on the transverse axis of the ellipse, and consequently it is equal to πA^2. Hence the area of the ellipse is equal to
$$\pi A^2 \times \frac{B}{A} = \pi AB.$$

Ex. 127. To find the area of a cycloid.

The differential equation of the cycloid is (Art. 130),
$$dx = \frac{y\,dy}{\sqrt{2ry - y^2}};$$
$$\therefore \int y\,dx = \int \frac{y^2 dy}{\sqrt{2ry - y^2}}.$$

And, by applying Formula E twice, we shall have
$$\int y\,dx = -\tfrac{1}{2}y\sqrt{2ry - y^2} - \tfrac{3}{2}r\sqrt{2ry - y^2} + \tfrac{3}{2}r^2 \text{ versin}^{-1}\left(\frac{y}{r}\right).$$

And making $y=0$ and $y=2r$, we have the area between those limits equal to

$\frac{3}{2}\pi r^2$; $\therefore$ whole area $= 3\pi r^2$.

That is, *the area of the cycloid is equal to three times the area of the generating circle.*

211.   AREA OF SPIRALS.

We have, by Art. 148, the differential of the area of any segment of a polar curve
$$ds = \tfrac{1}{2} r^2 d\theta.$$

Ex. 128. To find the area of the spiral of Archimedes. By Art. 134, the equation is
$$r = a\theta \text{ where } a = \frac{1}{2\pi};$$
$$\therefore s = \tfrac{1}{2} a^2 \int \theta^2 d\theta = \tfrac{1}{6} a^2 \theta^3 = \frac{\theta^3}{24\pi^2} + C.$$

Making $\theta = 0$ and $\theta = 2\pi$, we have the area included within the first spire $PMA = \tfrac{1}{3}\pi = $ *one third of the area of a circle whose radius is equal to the radius vector at the end of the first revolution.*

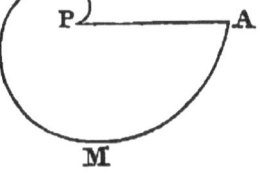

The radius vector in every subsequent revolution will pass over the area previously described. Hence, to find the area at the end of the nth revolution, we must integrate between the limits
$$\theta = (n-1)2\pi, \text{ and } \theta = 2n\pi,$$
which gives
$$\frac{n^3 - (n-1)^3}{3} \pi.$$

Ex. 129. To find the area of the hyperbolic or reciprocal spiral.

The equation, by Art. 136, is
$$r = \frac{a}{\theta};$$
$$\therefore s = \tfrac{1}{2} a^2 \int \theta^{-2} d\theta = -\tfrac{1}{2} \cdot \frac{a^2}{\theta}.$$

Ex. 130. To find the area of the logarithmic spiral. The equation, by Art. 135, is
$$\theta = \log. r;$$
$$\therefore d\theta = \frac{dr}{r},$$
and
$$s = \tfrac{1}{2} \int r dr = \tfrac{1}{4} r^2 + C.$$

If we make $r=0$, then $s=0$, and $C=0$;
$$\therefore s = \tfrac{1}{4}r^2.$$
That is, *the area of the Naperian logarithmic spiral is equal to one fourth of the square described on the radius vector.*

212. Area of Surfaces of Revolution.

The differential of the area of a surface of revolution is, by Art. 141,
$$dS = 2\pi y \sqrt{dx^2 + dy^2};$$
$$\therefore S = 2\pi \int y \sqrt{dx^2 + dy^2}.$$
If the curve were revolved about the axis of y, it would be necessary to change x into y and y into x.

Ex. 131. To find the convex surface of a cone.

Let $y = ax$ be the equation of the generating line; then $dy = adx$; $\therefore dy^2 = a^2 dx^2$,
and
$$dS = 2\pi y \sqrt{dx^2 + dy^2},$$
$$= 2\pi ax \sqrt{dx^2 + a^2 dx^2},$$
$$= 2\pi ax\, dx \sqrt{1 + a^2};$$
$$\therefore S = 2\pi a \sqrt{1 + a^2} \int x\, dx,$$
$$= \pi a \sqrt{1 + a^2} \cdot x^2 + C. \qquad (A)$$

But $a = \dfrac{y}{x}$;
$$\therefore a^2 = \frac{y^2}{x^2}, \text{ and } \sqrt{1+a^2} = \frac{1}{x}\sqrt{x^2 + y^2}.$$

Substituting these in (A), we have
$$S = \pi y \sqrt{x^2 + y^2} = 2\pi y \times \tfrac{1}{2}\sqrt{x^2 + y^2}.$$
But $\tfrac{1}{2}\sqrt{x^2 + y^2}$ = half the slant height of the cone, and $2\pi y$ = circumference of the base.

Therefore *the convex surface of a cone is equal to the circumference of the base multiplied by half the slant height.*

Cor. Taking the limits between x and $x+h$, we shall have the *convex surface of the frustum of a cone*
$$= \pi a \sqrt{1+a^2}\{(x+h)^2 - x^2\} = \{(x+h) + x\}h \cdot \pi a \sqrt{1+a^2}$$
$$= \pi\{a(x+h) + ax\}h\sqrt{1+a^2}.$$
Now $a(x+h)$ = the radius of the greater base,
ax = the radius of the lesser base,
and $h\sqrt{1+a^2}$ = the slant height of the frustum.

AREA OF SURFACES OF REVOLUTION. 285

Hence, to find the convex surface of the frustum of a cone, we have the following

RULE.

Multiply the sum of the radii of the upper and lower bases by the slant height, and that product by 3.1416.

Ex. 132. To find the convex surface of a cylinder.
Let $y=b$ be the equation of the generating line.
Then $\quad dy=0$,
and $\quad dS = 2\pi y \sqrt{dx^2 + dy^2} = 2\pi b\, dx;$
$\therefore S = 2\pi bx + C.$
When $x=0$, $C=0$, and $S=0$.
If we make $x=h=$ the height of the cylinder, we have
$$S = 2\pi b \cdot h.$$
That is, *the convex surface of a cylinder is equal to the circumference of its base multiplied by its altitude.*

Ex. 133. To find the surface of a sphere.
Let $y = \sqrt{r^2 - x^2}$ be the equation of the generating circle. Then
$$dy = -\frac{x\,dx}{y}, \text{ and } dy^2 = \frac{x^2 dx^2}{y^2};$$
$$\therefore dS = 2\pi y \sqrt{dx^2 + \frac{x^2 dx^2}{y^2}},$$
$$= 2\pi\, dx \sqrt{x^2 + y^2},$$
$$= 2\pi r\, dx,$$
or $\quad S = 2\pi r x + C.$
If $x=0$, $C=0$, and $S=0$.
If $x=r$, $S = 2\pi r^2 =$ surface of a hemisphere.
Hence $\quad 4\pi r^2 =$ whole surface.

That is, *the surface of a sphere is equal to four of its great circles, or equal to the convex surface of the circumscribing cylinder.*

Ex. 134. *To find the surface of a paraboloid.*
The equation of the generating parabola is
$$y^2 = 2px,$$
which, being differentiated, gives
$$dx = \frac{y\,dy}{p};$$
$$\therefore dx^2 = \frac{y^2 dy^2}{p^2}.$$

Hence $dS = \dfrac{2\pi}{p} y\,dy(p^2+y^2)^{\frac{1}{2}}$;

$\therefore S = \dfrac{2\pi}{p}\int y\,dy(p^2+y^2)^{\frac{1}{2}} = \dfrac{2\pi}{3p}(p^2+y^2)^{\frac{3}{2}} + C$,

by Art. 186.

If we make $y=0$, S will become equal to zero, and

$0 = \dfrac{2\pi p^2}{3} + C$; $\therefore C = -\tfrac{2}{3}\pi p^2$.

Now, integrating between the limits $y=0$ and $y=b$, we shall have

$$S = \dfrac{2\pi}{3p}\left[(p^2+b^2)^{\frac{3}{2}} - p^3\right].$$

Ex. 135. To find the surface of an ellipsoid formed by revolving an ellipse about the transverse axis.

The equation of the generating curve is

$$y^2 = \dfrac{B^2}{A^2}(A^2 - x^2),$$

whence $dy = -\dfrac{B}{A}\dfrac{x\,dx}{\sqrt{A^2-x^2}}$;

$\therefore dy^2 = \dfrac{B^2 x^2 dx^2}{A^2(A^2-x^2)}.$

Substituting this value of dy^2 and the value of y, which is $\dfrac{B}{A}(A^2-x^2)^{\frac{1}{2}}$, in the differential of the surface, we shall have, by a little reduction,

$$dS = \dfrac{2\pi B}{A}dx\left(A^2 - \dfrac{A^2-B^2}{A^2}x^2\right)^{\frac{1}{2}}.$$

If we put $\dfrac{A^2-B^2}{A^2} = e^2$,

we shall have $dS = \dfrac{2\pi Be}{A}dx\left(\dfrac{A^2}{e^2} - x^2\right)^{\frac{1}{2}}.$

Integrating by Formula B, we obtain

$$S = \dfrac{\pi Be}{A}\left[x\left(\dfrac{A^2}{e^2} - x^2\right)^{\frac{1}{2}} + \dfrac{A^2}{e^2}\sin^{-1}\dfrac{ex}{A}\right] + C.$$

If $x=0$, $C=0$, and $S=0$. And if we integrate between the limits $x=+A$ and $x=-A$, we shall have the whole surface

$$S = \dfrac{2\pi AB}{e}\{e\sqrt{1-e^2} + \sin^{-1} e\}.$$

Ex. 136. To find the surface of an oblate spheroid.

In order to determine the equation to this, we merely change A into B, and we have
$$y^2 = \frac{A^2}{B^2}(B^2 - x^2),$$
from which the surface of the oblate spheroid may be found in the same manner as the last, by integrating between the limits $x = +B$ and $x = -B$.

Ex. 137. To find the surface described by the revolution of the cycloid about its base.

The differential equation of the cycloid (Art. 130) is
$$dx = \frac{y\,dy}{\sqrt{2ry - y^2}};$$
$$\therefore dS = \frac{2\pi(2r)^{\frac{1}{2}} y^{\frac{3}{2}} dy}{\sqrt{2ry - y^2}},$$
and
$$S = 2\pi(2r)^{\frac{1}{2}} \int \frac{y^{\frac{3}{2}} dy}{\sqrt{2ry - y^2}}.$$

Applying Formula E (Art. 200), we have
$$S = 2\pi(2r)^{\frac{1}{2}} \left[-\tfrac{2}{3} y^{\frac{1}{2}} (2ry - y^2)^{\frac{1}{2}} + \tfrac{4}{3} r \int \frac{y^{\frac{1}{2}} dy}{\sqrt{2ry - y^2}} \right].$$

But, by Art. 186,
$$\int \frac{y^{\frac{1}{2}} dy}{\sqrt{2ry - y^2}} = \int \frac{dy}{\sqrt{2r - y}} = \int dy (2r - y)^{-\frac{1}{2}} = -2(2r - y)^{\frac{1}{2}}.$$

Hence $S = 2\pi(2r)^{\frac{1}{2}} \left[-\tfrac{2}{3} y^{\frac{1}{2}} (2ry - y^2)^{\frac{1}{2}} - \tfrac{8}{3} r (2r - y)^{\frac{1}{2}} \right] + C.$

If we estimate the surface from a plane passing through the centre B, we shall have $y = 2r$, $C = 0$, and $S = 0$. If we then integrate between the limits $y = 0$ and $y = 2r$, we shall have

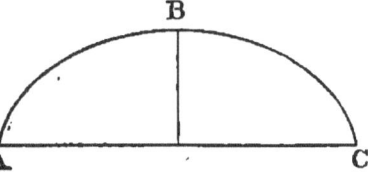

half the surface $= -\tfrac{32}{3} \pi r^2$;
hence surface $= -\tfrac{64}{3} \pi r^2$.

That is, *the surface described by the cycloid revolved around its base is equal to 64 thirds of the generating circle.*

CUBATURE OF SOLIDS OF REVOLUTION.

213. The *cubature* of a solid is the expression of its volume or capacity.

By Art. 142, the differential of a solid of revolution is
$$dV = \pi y^2 dx; \quad \therefore V = \pi \int y^2 dx, \tag{1}$$
where x and y are the co-ordinates of any point of the generating curve, and the axis of x is the axis of revolution. But if the axis of y were the axis of revolution, we should have for the differential of the volume
$$dV = \pi x^2 dy; \quad \therefore V = \pi \int x^2 dx. \tag{2}$$
Hence, to find the volume of a solid formed by revolving a line around the axis of x, we have the following

RULE.

From the equation of the line find the value of y^2; multiply this by πdx, and integrate between the proper limits.

Ex. 138. To find the volume of a cone.

Let $y = ax$ be the equation of the generating line. Then
$$\pi y^2 dx = \pi a^2 x^2 dx; \tag{2}$$
hence
$$V = \pi a^2 \int x^2 dx = \tfrac{1}{3}\pi a^2 x^3 + C; \tag{3}$$
and taking the limits between $x = 0$ and $x = h$, we have
$$V = \tfrac{1}{3}\pi a^2 h^3.$$

But $\quad a = \dfrac{y}{x}.$

Substituting this in (Eq. 3) above, we obtain
$$V = \tfrac{1}{3}\pi y^2 h = \pi y^2 \times \tfrac{1}{3}h, \text{ by making } x = h.$$

That is, *the volume of a cone is equal to the area of its base into one third of its altitude.*

Cor. If we integrate (Eq. 3) between the limits x and $x + h$, we shall have the volume of the frustum of a cone,
$$V = \tfrac{1}{3}\pi a^2 \{(x+h)^3 - x^3\},$$
$$= \tfrac{1}{3}\pi h \{3a^2 x^2 + 3a^2 xh + a^2 h^2\},$$
$$= \tfrac{1}{3}\pi h \{(ax)^2 + (ax)a(x+h) + [a(x+h)]^2\}.$$

But $\quad ax = y = r =$ radius of the upper base,
and $\quad a(x+h) = R =$ radius of the lower base;

$\therefore$ vol. of frustum of a cone $= \tfrac{1}{3}\pi h(r^2 + rR + R^2).$

Ex. 139. To find the volume of a prolate spheroid formed by revolving an ellipse about its transverse axis.

The equation of the ellipse is

$$y^2 = \frac{B^2}{A^2}(A^2 - x^2);$$

$$\therefore \pi y^2 dx = \pi \frac{B^2}{A^2}(A^2 - x^2) dx,$$

and
$$V = \pi \frac{B^2}{A^2} \int (A^2 - x^2) dx,$$

$$= \pi \frac{B^2}{A^2}(A^2 x - \tfrac{1}{3}x^3) + C,$$

which is the volume of the spheroidal segment.

If we estimate the volume from a plane passing through the centre perpendicular to the axis of revolution, we shall have, when $x = 0$, $V = 0$, and therefore $C = 0$.

Integrating between the limits $x = +A$ and $x = -A$, we obtain

$$V = \tfrac{2}{3}\pi B^2 \times 2A,$$

where πB^2 is the area of a mid section, and $2A$ is the transverse axis. Hence *the volume of a prolate spheroid is equal to two thirds of the circumscribing cylinder.*

Cor. If we make $A = B$, we shall obtain the volume of a sphere,

$$\tfrac{4}{3}\pi r^3 = \tfrac{1}{6}\pi D^2,$$

where r is the radius and D the diameter of the sphere.

Ex. 140. Find the volume of the sphere from the equation $y^2 = r^2 - x^2$.

Ex. 141. Find the volume of the sphere from the equation $y^2 = 2rx - x^2$.

Ex. 142. Find the volume of an oblate spheroid formed by revolving the curve about the axis of y.

The equation of the curve is

$$x^2 = \frac{A^2}{B^2}(B^2 - y^2);$$

$$\therefore \pi x^2 dy = \pi \frac{A^2}{B^2}(B^2 - y^2) dy,$$

and
$$V = \pi \frac{A^2}{B^2}(B^2 y - \tfrac{1}{3}y^3) + C.$$

Integrating between $y = +B$ and $y = -B$, we obtain
$$V = \tfrac{2}{3}\pi A^2 \times 2B.$$

If we compare the volumes of the prolate spheroid and the oblate spheroid, we find the

prolate spheroid : oblate spheroid :: B : A.

Ex. 143. Find the volume of a paraboloid.
The equation of the generating curve is
$$y^2 = 2px;$$
$$\therefore \pi y^2 dx = 2\pi px dx,$$
and $\quad V = \pi p x^2.$

But $px^2 = \tfrac{1}{2}y^2 x$ from the equation of the generating curve.
$$\therefore V = \tfrac{1}{2}\pi y^2 \cdot x = \pi y^2 \cdot \tfrac{1}{2}x.$$
That is, *the volume of a paraboloid is equal to one half of a cylinder of equal base and altitude.*

Ex. 144. Find the volume of a frustum of a paraboloid.
Let R and r denote the radii of the lower and upper bases respectively, and h the altitude of the frustum. Then
$$y^2 = 2px,$$
and $\quad \pi y^2 dx = 2\pi px dx.$

Integrating this between the limits $(x+h)$ and x, we have
$$\text{vol.} = \pi p\{(x+h)^2 - x^2\} = \pi p(2x+h)h,$$
$$= \pi p\{(x+h)+x\}h,$$
$$= \tfrac{1}{2}\pi h\{2p(x+h)+2px\}.$$
But, from the property of the parabola,
$$2px = r^2, \text{ and } 2p(x+h) = R^2;$$
$$\therefore \text{vol.} = \tfrac{1}{2}\pi(R^2 + r^2)h.$$
That is, *the volume of a frustum of a paraboloid is equal to half the sum of the areas of its two bases multiplied by the altitude of the frustum.*

Ex. 145. Find the volume of the solid generated by the revolution of a cycloid around its base.
The differential equation of the cycloid (Art. 130) is
$$dx = \frac{y\,dy}{\sqrt{2ry - y^2}};$$
$$\therefore \pi y^2 dx = \frac{\pi y^3 dy}{\sqrt{2ry - y^2}},$$
which, being integrated by Formula E and the final integral $\int \frac{dy}{\sqrt{2ry - y^2}}$, by Art. 189, Ex. 52, we shall find *the volume equal to five eighths of the circumscribing cylinder.*

214. There is another method by which volumes of revolution may be conceived to be generated, that is, by the

motion of a curve parallel to its own plane, and varying in magnitude according to a fixed law. Thus the volume of revolution may be considered as generated by the motion of a circle whose centre is in the axis of x, its plane being always perpendicular to that axis, and its radius varying according to the law of the variation of the ordinates of the meridian curve; and, viewing the generation as effected in this manner, we may say that *the differential of the volume is equal to the area of the generating circle multiplied by the differential of the axis*, as before.

This theorem is true whatever be the form of the generating area, provided only that, as in the case of the circle, it varies agreeably to some law dependent on the equation of the directrix; or, in other words, provided we can always express this in general terms as an invariable function of the general co-ordinates of the directrix.

Ex. 146. Let us give an example of its application to the solid called *a circular groin*.

The generating area in this case is a *square*, and the meridian curve is a semicircle passing through the middle points of its opposite sides.

Taking the diameter perpendicular to the moving plane as the axis of x, and the vertex of the groin as the origin of rectangular co-ordinates, we shall have for half the side of the square

$$y = \sqrt{2rx - x^2}.$$

Hence the side of the square

$$= 2\sqrt{2rx - x^2},$$

and its area $\qquad = 4(2rx - x^2).$

Multiplying this by dx gives

$$dV = 4(2rx - x^2)dx;$$
$$\therefore V = 4rx^2 - \tfrac{4}{3}x^3 + C.$$

When $x=0$, $C=0$. Integrating between $x=0$ and $x=r$, we shall have

$$V = \tfrac{8}{3}r^3.$$

Ex. 147. Find the volume of an ellipsoid, the equation being $\quad A^2B^2z^2 + B^2C^2x^2 + A^2C^2y^2 = A^2B^2C^2.$

At any distance, z, from the plane xy and parallel thereto, we have

$$A^2y^2 + B^2x^2 = \frac{A^2B^2(C^2-z^2)}{C^2}$$

for the equation of the section.

Let us consider this as the generating ellipse, and take for its semi-axes the variable quantities a and β.

Hence
$$a = \frac{A(C^2-z^2)^{\frac{1}{2}}}{C},$$

$$\beta = \frac{B(C^2-z^2)^{\frac{1}{2}}}{C};$$

and consequently its area (Ex. 126) is
$$\pi a\beta = \frac{\pi AB(C^2-z^2)}{C^2}.$$

Now, multiplying this by dz and integrating, we have
$$V = \frac{\pi AB}{C^2}(C^2 z - \tfrac{1}{3}z^3) + C.$$

Taking the limits between $z=0$ and $z=+C$, we have, for one half the volume, $\tfrac{2}{3}\pi ABC$.

$$\therefore V = \tfrac{4}{3}\pi ABC = \text{whole volume}.$$

215. The formulas and examples which we have now given for the quadrature of curved surfaces and the cubature of their volumes will be found sufficient for nearly all practical purposes, inasmuch as the curved surfaces employed in the arts are almost invariably surfaces of revolution. In order, however, to make the subject more complete, we shall investigate general expressions for the volume and surface of any body that can be represented by an equation, which will require

DOUBLE INTEGRALS.

216. Let us take a solid bounded by three rectangular planes and by any surface whose equation is given.

Assume A as the origin, and AX, AY, AZ, the three co-ordinate axes. On AX take any distance, $AB=x$, and through B pass a plane NPE parallel to the plane of yz; also on AY take any distance, $AI=y$, and through I pass a plane MPH parallel to the plane xz; these two planes will have the line PS as their intersection, which we may put equal to z.

It is evident that the solid whose base is AISB may

be extended toward XH without changing the value of y, or toward YE without altering the value of x. Hence x and y may be considered as independent variables.

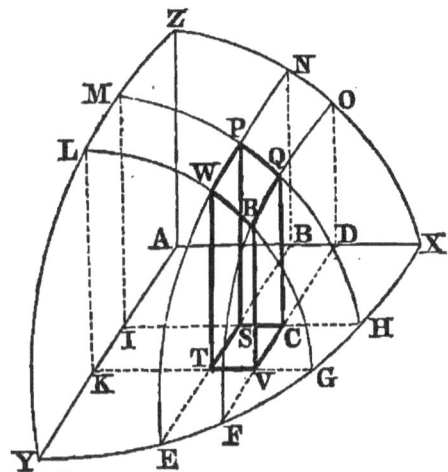

Give now to x an increment, $BD = h$ (y remaining constant), and through D pass a plane OQF parallel to the plane of yz, the solid whose base is AISB will be augmented by the solid whose base is BSCD; or, if we give to y an increment, $IK = k$ (x remaining constant), and pass the plane LWG through k parallel to the plane of xz, the solid whose base is AISB will be augmented by the solid whose base is ISTK; but if we give to x and y these increments simultaneously, and pass the planes OQF and LWG respectively parallel to the co-ordinate planes, the solid thus cut off will differ from the solid whose base is AISB by the two solids before named and the solid whose base is STVC.

Let $u =$ the volume of the solid whose base is AISB: we shall have u a function of x and y. Hence, if we take the sum of the increments of $f(x, y)$ under the supposition that one at a time varies from the increment of that function when the variables have received simultaneous increments, the remainder will be the solid whose base is the rectangle STVC.

Therefore, Art. 64,

$$\text{vol. of solid STVC} \ldots \text{P} = \frac{d^2 u}{dx\,dy} hk + \frac{d^3 u}{dx^2 dy} \frac{h^2 k}{1 \cdot 2} +, \text{etc.};$$

and by dividing by hk, we have

$$\frac{\text{vol. of solid}}{hk} = \frac{d^2 u}{dx\,dy} + \frac{d^3 u}{dx^2 dy} h +, \text{etc.}$$

In the first member of this equation,
$hk =$ rectangle STVC;

and the volume divided by the base is equal to the altitude of the prism $=z$; therefore, passing to the limit by making h and k each equal to zero, we have

$$z = \frac{d^2u}{dxdy},$$

or $\qquad d^2u = zdxdy.$
Hence $\qquad u = z\int dx \int dy.$

From which we derive the following

Rule.

From the equation of the given surface find the value of z; multiply this by dx, y being considered constant, and integrate between the limits $x=0$, and $x=$ the value it has in the equation of the surface when z equals zero. Multiply this corrected integral by dy, and integrate between the limits $y=0$, and the value which y assumes when x and z are both equal to zero in the equation of the given surface. The final integral will give the volume contained in the first angle.

We have given this rule for the sake of convenience, but we may find the value of z from the equation of the given surface; multiply this by the differential of either of the other variables, and integrate according to the rule between the proper limits; then multiply this corrected integral by the differential of the other variable; integrate between the limits, and this final integral will be the volume contained in the first angle of the three co-ordinate planes.

Ex. 148. Find the volume of a sphere.

Let the centre of the sphere be at the origin of co-ordinates. Then, if R denote the radius, we have

(1) $\qquad x^2 + y^2 + z^2 = R^2; \therefore z = (R^2 - x^2 - y^2)^{\frac{1}{2}},$

and consequently

$$z \int dx \int dy = \int dx \int dy \{(R^2 - x^2) - y^2\}^{\frac{1}{2}}.$$

If we now consider x constant, and integrate with respect to y by Formula B and Art. 189, we shall have

$$\int dy \{(R^2 - x^2) - y^2\}^{\frac{1}{2}} =$$

$$\tfrac{1}{2}y\{(R^2 - x^2) - y^2\}^{\frac{1}{2}} + \tfrac{1}{2}(R^2 - x^2) \sin^{-1} \frac{x}{\sqrt{R^2 - x^2}}.$$

Taking the limits between $y=0$ and $y=\sqrt{R^2-x^2}$, obtained in (1) under the supposition that $z=0$, we have

$$\int dy\{(R^2-x^2)-y^2\}^{\frac{1}{2}} = \frac{\pi}{4}(R^2-x^2).$$

Multiply this by dx, and integrate

$$\int dx \int dy\{(R^2-x^2)-y^2\}^{\frac{1}{2}} = \frac{\pi}{4}\int (R^2-x^2)dx,$$

$$= \frac{\pi}{4}(R^2 x - \tfrac{1}{3}x^3) + C.$$

When $x=0$, the integral is equal to zero; when $x=R$, the integral is $=\tfrac{1}{6}\pi R^3$, which represents that part of the sphere which is contained in the first angle of the co-ordinate planes, or *one eighth of the whole volume*. Hence vol. of the sphere $=\tfrac{4}{3}\pi R^3 = \tfrac{1}{6}D^3\pi$.

The first integral $z \int dy$ expresses the area of the section NWE parallel to the co-ordinate plane yz for any abscissa $AB = x$, and to obtain this area we must integrate between the limits $y=0$ and $y=BE$. But, since E is in the co-ordinate plane yx, we must have for this point $z=0$, and the equation of the surface gives

$$BE = \sqrt{R^2-x^2} = y.$$

Hence we take the limits of y between $y=0$ and $y=\sqrt{R^2-x^2}$, as above.

Ex. 149. Find the volume of the solid formed by three rectangular planes and the plane whose equation is

$$x+2y+3z-2=0.$$

Ans. Vol. $=\tfrac{2}{9}$.

Ex. 150. Find the volume of the general ellipsoid, the equation being $\dfrac{x^2}{a^2}+\dfrac{y^2}{b^2}+\dfrac{z^2}{c^2}=1.$

Ans. $V = \tfrac{4}{3}\pi abc$.

THE VOLUME A FUNCTION OF THREE VARIABLES.

217. If the volume be supposed to be a function of three variables, x, y, and z, we shall have

$$\frac{\text{solid}}{hkl} = \frac{d^3u}{dxdydz}+, \text{ etc.};$$

and, passing to the limit, the solid becomes a prism whose base is hk and its altitude l, and therefore the volume $=hkl$. Hence we obtain

$$\frac{d^3u}{dx\,dy\,dz}=1,$$

or $u = \iiint dx\,dy\,dz.$

218. If S denote the surface of the solid, we shall find, in a manner exactly similar to the above,

$$\frac{\text{surface}}{hk} = \frac{d^2 S}{dx\,dy} +, \text{ etc.}$$

Passing to the limit, PQRW becomes coincident with a tangent plane, and is equal to the base SCVT × secant of the inclination of the tangent plane to the plane of xy

$$= hk \sec. < = hk \sqrt{1 + \frac{dz^2}{dx^2} + \frac{dz^2}{dy^2}};$$

$$\therefore S = \iint dx\,dy \sqrt{1 + \frac{dz^2}{dx^2} + \frac{dz^2}{dy^2}}.$$

To apply this formula, let us take as an example,

Ex. 151. To determine the surface of a sphere from the equation

$$x^2 + y^2 + z^2 = r^2.$$

By differentiating, we have

$$\frac{dz}{dx} = -\frac{x}{z}, \text{ and } \frac{dz}{dy} = -\frac{y}{z}.$$

Substitute these in the differential of the surface, and we have

$$S = \iint dx\,dy \sqrt{1 + \frac{x^2}{z^2} + \frac{y^2}{z^2}},$$

$$= \iint dx\,dy \frac{\sqrt{x^2 + y^2 + z^2}}{z},$$

$$= \iint dx\,dy \frac{r}{\sqrt{r^2 - y^2 - z^2}};$$

$$\therefore S = r \int_{-r}^{+r} dy \int_{-\sqrt{r^2-y^2}}^{+\sqrt{r^2-y^2}} dx\,(r^2 - y^2 - x^2)^{-\frac{1}{2}},$$

whole surface $= 4\pi r^2$.

219. Having conducted the student through the elements of the Differential and Integral Calculus on the principles first established by Sir Isaac Newton, we shall proceed to explain, briefly, the Theory of Infinitesimals, as originated by the profound mathematician Leibnitz.

Definition.—An *infinite quantity* is a quantity which can not be increased.

In the expression $x+a$, if we suppose x to be infinite, we must suppress a, otherwise it would appear as if x were increased by a, which is contrary to the definition.

For, let
$$\frac{1}{a}+\frac{1}{x}=b;\qquad(1)$$
then $\qquad x+a=abx.\qquad(2)$

Now, if we suppose x to be infinite, (Eq. 1) will become
$ab=1$, since $\frac{1}{\infty}=0$, Art. 7.

Substituting this in (2), we obtain
$$x+a=x,\qquad(3)$$
which shows that when x is infinite, $x+a$ is equal to x.

The quantity a, in comparison with which x is infinite, is called an *infinitesimal*, or an infinitely small quantity in reference to x.

As we are now considering only the relative values of quantities, the preceding will hold good even when x has a *finite* value, provided only that a be infinitely small in comparison with x.

Although two quantities be infinitely small, it does not follow that their ratio will be *zero*. We found in Art. 36, by making $h=0$, that
$$\frac{du}{dx}=a,$$
where du and dx were made infinitely small. The ratio is here determined $=a$. Now, if we assign any value whatever to the symbol du, we shall find a corresponding value for the other symbol, $dx;$ or, if we assign a value to dx, a corresponding value will be obtained for du. These truths are exemplified in Ex. 29, page 124, and elsewhere.

When a quantity, x, is infinitely small relatively to a finite magnitude, a, the square of x is infinitely small relatively to x. For the proportion
$$1:x::x:x^2$$
shows that x^2 is involved in x as often as x is involved in unity; that is, an infinite number of times.

We may show in the same manner by the proportion
$$x:x^2::x^2:x^3,$$

N 2

that if x^2 is infinitely small relatively to x, the quantity x^3 must be infinitely small relatively to x^2.

Infinitesimals are, therefore, according to this view of the subject, divided into different orders: thus, in the foregoing examples, x is an infinitesimal of the first order, x^2 is an infinitesimal of the second order, x^3 is an infinitesimal of the third order, and so on.

Since an infinitesimal of the first order must be disregarded when connected with a finite quantity which it can not increase, so an infinitesimal of the second order must be disregarded when connected with an infinitesimal of the first order, and so on.

The product of two infinitesimals of the first order is an infinitesimal of the *second* order.

Thus, if
$$1:x::y:xy,$$
xy will be infinitely small relatively to y, since x is infinitely small relative to unity. Hence xy will be an infinitesimal of the second order.

In like manner we might show that the product of three infinitesimals of the first order is an infinitesimal of the *third order*.

The Theory of Infinitesimals has been considered by writers on the Calculus as less philosophical than that of the Method of Limits. But its results are *precisely the same*, while the labor it requires to obtain those results is by no means so severe. This gives it great advantage over the other in the discussion of astronomical and physical problems, and is therefore used by authors in treating those subjects.

We shall give the reader a few examples to illustrate that system.

(1.) Let
$$u = f(x) = ax. \qquad (1)$$

Now, let x be increased by an infinitely small quantity, which we will represent by dx, called the differential of x, and let the corresponding infinitely small increment of u be denoted by du. Then, when x becomes $x+dx$, u will become $u+du$. Substituting these in (1), we have
$$u+du = ax+adx. \qquad (2)$$

Subtracting (1) from (2), we obtain
$$du = adx,$$
or
$$\frac{du}{dx} = a. \text{ (The same as Art. 36.)}$$

THE THEORY OF INFINITESIMALS. 299

(2.) Let $u = ax^2$;
then $u + du = ax^2 + 2ax dx + a dx^2$;
$\therefore du = 2ax dx + a dx^2$.

But dx^2, being an infinitesimal of the second order, can not increase $2ax dx$, and must therefore be rejected.
Hence $du = 2ax dx$,
or $\dfrac{du}{dx} = 2ax$. (The same as Art. 38.)

(3.) Let $u = yz$, where y and z are both functions of x.
Then $u + du = (y + dy)(z + dz)$,
$= yz + ydz + zdy + dy.dz$;
$\therefore du = ydz + zdy + dy.dz$.

But $dy.dz$ is an infinitesimal of the second order, and must be suppressed.
$\therefore du = ydz + zdy$. (The same as Art. 43.)

(4.) Determine the differential of the sine of an arc.
Let $u = \sin. x$;
then $u + du = \sin. (x + dx)$,
$= \sin. x \cos. dx + \cos. x \sin. dx$.

But when the arc is infinitely small, as dx,
$\cos. dx = 1$,
and $\sin. x = dx$;
$\therefore u + du = \sin. x + \cos. x dx$.
Hence $du = \cos. x dx$. (As Art. 74.)

(5.) Determine the differential of the length of an arc referred to rectangular axes.

Let z denote the length of the arc, and x and y the co-ordinates of any point of the curve. Then will dx and dy represent the sides, and dz the hypothenuse of an indefinitely small right angled triangle. Hence

$$dz = \sqrt{dx^2 + dy^2}. \text{ (As Art. 139.)}$$

(6.) Determine the differential of the area of a curve referred to rectangular axes.

If s denote the area, x and y the co-ordinates of any point, then the differential of the area is a trapezoid whose parallel sides are y and $y + dy$, and the distance between them $= dx$. Therefore

$$ds = \tfrac{1}{2}(y + y + dy)dx = ydx + \tfrac{1}{2}dy.dx.$$

But $dy.dx$ being an infinitesimal of the second order, we shall have
$du = ydx$. (As Art. 140.)

(7.) Determine the differential of the surface and volume of a solid of revolution.

If we suppose the solid to be composed of an infinite number of laminæ or indefinitely thin plates perpendicular to the axis, then each of these *laminæ* will be a circle. Now the circumferences of these circles make up the whole surface of the solid; and if we let S denote the surface, V the volume, and x and y the co-ordinates of any point of the generating curve, we shall have $2\pi y$ for the length of the circumference of any one of these circles, and the differential of the arc of the generating curve for its breadth. Hence

$$dS = 2\pi y \sqrt{dx^2 + dy^2}.$$ (The same as Art. 141.)

Again: The volumes of these *laminæ* evidently make up the whole volume of the solid; but each of these *laminæ* is a cylinder the area of whose base is πy^2, and its altitude is the differential of x.

Hence $dV = \pi y^2 dx$. (The same as Art. 142.)

(8.) For the differential of the volume of a parallelopipedon we shall have

$$d^3V = dx\,dy\,dz;$$
$$\therefore V = \int dx \int dy \int dz = \text{vol. of any solid},$$

subject, however, to the proper limits.

Miscellaneous Examples.

Ex. 152. Determine the co-ordinates of the point which divides a given line into two parts in the ratio of m to n.

Ex. 153. Express the area of a triangle in terms of the co-ordinates of its angular points.

Ex. 154. Determine the co-ordinates of the point of intersection of the lines $y = -x + 1$ and $y = x + 2$; also the angle contained between them.

Ex. 155. Determine the angle contained between the lines $y = -\frac{x}{3} + \frac{1}{2}$ and $y = \frac{x}{2} - \frac{1}{2}$.

Ex. 156. Determine the length of the perpendicular drawn from the point $(1, -2)$ to the line whose equation is $y = -x + 3$.

Ans. $2\sqrt{2}$.

Ex. 157. Construct the circles whose equations are
$$x^2+y^2-8y+4x-5=0.$$
Ex. 158. $x^2+y^2+4y-4x-1=0.$
Ex. 159. $x^2+y^2-3y+6x-1=0.$

Ex. 160. Determine the equation to a straight line passing through the point $x=1$, $y=3$, and making an angle of $30°$ with the line $2y-x+1=0$.

Ex. 161. Determine the equation to the line which cuts the axis of X at a distance c from the origin, and makes an angle of $45°$ with the line $\dfrac{x}{a}-\dfrac{y}{b}=1$.

Ex. 162. Determine the distance of the intersection of the two lines $2x-3y+5=0$ and $3x+4y=0$ from the line $y=2x+1$.

Ex. 163. Determine the tangent of the angle included between the lines $2x+3y+4=0$ and $3x+4y+3=0$.

Ex. 164. Determine the equation to the line that bisects the angle made by the lines $5y-2x=0$ and $3y+4x=12$.

Ex. 165. The equations of two lines are $y=-2x+6$ and $y=a'x-4$: what will the equation of the second line become when it is perpendicular to the first?

Ex. 166. Given the line $y=-3x+4$, and the point $x=7$, $y=3$; determine the equation of a line passing through the point and perpendicular to the given line; also determine the co-ordinates of the point of intersection of the two lines, and find the length of the perpendicular.

Ex. 167. Determine the polar co-ordinates of the following points, $x=1$, $y=1$; $x=-1$, $y=1$; $x=-1$, $y=2$.

Ex. 168. Determine the rectangular co-ordinates of the following points, $\theta=\dfrac{\pi}{4}$, $r=5$; $\theta=\dfrac{\pi}{4}$, $r=-7$; $\theta=-\dfrac{\pi}{6}$, $r=-3$.

Ex. 169. Determine the area of the triangle whose vertices are the points $x=1$, $y=1$; $x=-1$, $y=1$; $x=-1$, $y=2$.

Ex. 170. Determine the area of the triangle formed by the lines $x+2y-5=0$, $2x+y-7=0$, and $y-x-1=0$.

Ex. 171. Determine the area of the triangle formed by the axes and the line $4x+3y=12$.

Ex. 172. P and P' are two points, and A is the origin; express the area of the triangle PAP' in terms of the co-ordinates of P and P', and also in terms of the polar co-ordinates of P and P'.

Ex. 173. Construct the line whose polar equation is $1 = \cos.\left(\theta - \dfrac{\pi}{4}\right)$.

Ex. 174. Determine the centre and radius of the circle whose equation is $x^2 + y^2 + 4x - 6y - 3 = 0$.

Ex. 175. Determine the equation to a straight line passing through the centres of the circles $x^2 + 4x + y^2 + 6y = 3$, and $x^2 + y^2 + 2y = 0$.

Ex. 176. Determine the equation to the circle passing through the three points (1, 2), (1, 3), and (2, 5).

Ex. 177. Determine the equation to a tangent to the circle $x^2 + y^2 - 3x + 4y = 0$, when the tangent passes through the origin.

Ex. 178. Determine the equation to a circle whose centre is at the origin, and which touches the line $y = 3x + 2$.

Ex. 179. Determine the equation to a circle passing through the origin and cutting off positively on the axes of X and Y the distances h and k respectively.

Ex. 180. Determine the intersection of the circle $x^2 + y^2 = 25$ with the line $y + x = -1$.

Ex. 181. Determine the equation to and length of the common chord of the two circles $x^2 - 4x + y^2 - 2y - 11 = 0$ and $x^2 + 6x + y^2 + 4y - 3 = 0$.

Ex. 182. Determine the point in the parabola the tangent at which makes an angle of 30° with the axis of X.

Ex. 183. Determine the point in the parabola the tangent at which cuts off equal distances on the axes.

Ex. 184. Determine the co-ordinates of the points where the line $y = ax + b$ meets the parabola $y^2 = 2px$, and then find the co-ordinates of the middle point of the chord.

Ex. 185. At the point (x'', y'') a normal is drawn; required the other point of intersection of this with the curve; also the length of the intercepted chord.

Ex. 186. Determine the polar equation to the parabola referred to the foot of the directrix as the pole, with the axis of the parabola as the inititial line.

Ex. 187. An ellipse and hyperbola have the same foci; show that their tangents at either point of intersection are perpendicular to each other.

Ex. 188. The equation of a plane being
$$Ax+By+Cz+D=0,$$
prove that the area of the triangle intercepted on this plane by the three rectangular co-ordinate planes is
$$\frac{D^2}{2ABC}\sqrt{A^2+B^2+C^2}.$$

Ex. 189. Determine the points of intersection of the two ellipses
$$7y^2+4x^2=28,$$
$$6y^2+5x^2=30,$$
referred to the same axes of co-ordinates, and determine also the angle they make with each other at those points.

Ans. $x=\pm\sqrt{\frac{42}{11}}$,
$y=\pm 2\sqrt{\frac{5}{11}}$,
and the angle $=10°\ 44'\ 43''$.

Ex. 190. The semi-axes of two ellipses are 2 and 1, and 5 and 3: it is required to place them, with their transverse axes, on the same straight line, so that they may intersect each other at right angles.

Ans. Distance between their centres $=6.149$.
Abscissa of point of intersection $=4.838$.

Ex. 191. If from any point on an ellipse a straight line be drawn to the centre, making an angle θ with the normal, and if ϕ be the angle which the normal makes with the transverse axis, show that
$$\tan.\ \theta=\frac{(a^2-b^2)\tan.\ \phi}{a^2+b^2\tan.^2\phi},$$
where a and b are the semi-axes.

Ex. 192. Differentiate $u=\sqrt{x+\sqrt{1+x^2}}$.

Ans. $\dfrac{du}{dx}=\dfrac{\sqrt{x+\sqrt{1+x}}}{2\sqrt{1+x^2}}$.

Ex. 193. Differentiate
$$u=\sqrt{a+x+\sqrt{a+x+\sqrt{a+x}}}+,\text{ etc., }ad\ inf.$$
Ans. $\dfrac{du}{dx}=\dfrac{1}{2u-1}$.

Ex. 194. Differentiate $u = 1 + \cfrac{x}{1 + \cfrac{x}{1 + \cfrac{x}{1+, \text{etc.}}}}$, ad inf.

$$\text{Ans. } \frac{du}{dx} = \frac{1}{\sqrt{4x+1}}.$$

Ex. 195. Differentiate the implicit function
$2ux + au^2 - bx^2 = 0$.

$$\text{Ans. } \frac{du}{dx} = \frac{u}{x}.$$

Ex. 196. Differentiate $u = \sin^{-1}\left(\dfrac{1-x^2}{1+x^2}\right)$.

$$\text{Ans. } \frac{du}{dx} = -\frac{2}{1+x^2}.$$

Ex. 197. Differentiate $u = \sin^{-1}\left(\dfrac{x^2-a^2}{b^2-a^2}\right)^{\frac{1}{2}}$.

$$\text{Ans. } \frac{du}{dx} = \frac{x}{\sqrt{(x^2-a^2)(b^2-x^2)}}.$$

Ex. 198. Differentiate $u = \tan^{-1}\left(\dfrac{1-x}{1+x}\right)^{\frac{1}{2}}$.

$$\text{Ans. } \frac{du}{dx} = -\frac{1}{2\sqrt{1-x^2}}.$$

Ex. 199. Differentiate $u = l \cdot \dfrac{\sqrt{a} + \sqrt{x}}{\sqrt{a} - \sqrt{x}}$.

$$\text{Ans. } \frac{du}{dx} = \frac{\sqrt{a}}{\sqrt{x}(a-x)}.$$

Ex. 200. Differentiate $u = xe^{\tan^{-1}x}$.

$$\text{Ans. } \frac{du}{dx} = \frac{e^{\tan^{-1}x}(1+x+x^2)}{1+x^2}.$$

Ex. 201. Find the equations to the asymptotes of the curve whose equation is
$y^4 = x^4 - 2bx^2y$.

$$\text{Ans. } y = \pm x - \frac{b}{2}.$$

Ex. 202. If $y = \dfrac{a^2 x}{a^2 + x^2}$, show that the curve cuts the axis of x at the origin, making an angle of 45°, and that the axis of x is an asymptote to the two infinite branches.

Ex. 203. If $y=x\cdot\sqrt{\dfrac{a^2+x^2}{a^2-x^2}}$, show that the branches of the curve pass through the origin, and are contained between two asymptotes perpendicular to the axis of x.

Ex. 204. Show that the curve whose equation is
$$(y^2+x^2)^3=4a^2x^2y^2$$
has a quadruple point at the origin, and that it has four ovals, the axis of each being equal to a.

Ex. 205. Determine the radius of curvature to the rectangular hyperbola, its equation being
$$xy=a^2.$$

$$Ans.\ \mathrm{R}=-\frac{(x^2+y^2)^{\frac{3}{2}}}{2a^2}.$$

Ex. 206. Determine the radius of curvature to the hyperbola, and also the equation of its evolute.

$$Ans.\ \mathrm{R}=\frac{(e^2x^2-a^2)^{\frac{3}{2}}}{ab}.$$

Eq. to evolute $(a\alpha)^{\frac{2}{3}}-(b\beta)^{\frac{2}{3}}=c^{\frac{4}{3}}$.

Ex. 207. Determine the co-ordinates of the point at which a tangent line is perpendicular to the axis of x on a curve whose equation is
$$y-2=(x-1)(x-2)^{\frac{1}{2}}.$$
$Ans.\ x=2$, and $y=2$.

Ex. 208. If a tangent be drawn to the curve whose equation is $y^{\frac{2}{3}}+x^{\frac{2}{3}}=a^{\frac{2}{3}}$, show that the part of the tangent which is intercepted between the axes is equal to a.

Ex. 209. Find the evolute to the curve whose equation is
$$y^{\frac{2}{3}}+x^{\frac{2}{3}}=a^{\frac{2}{3}}.$$

$$Ans.\ (a+\beta)^{\frac{2}{3}}+(a-\beta)^{\frac{2}{3}}=a^{\frac{2}{3}}.$$

Ex. 210. Let $z=\dfrac{h}{r}\sqrt{x^2+y^2}$ be the equation of a cone where h denotes the altitude, and r the radius of the base; to find the volume.

$Ans.\ \frac{1}{3}\pi r^2 h.$

Ex. 211. The axes of two equal cylinders intersect at right angles; to determine the volume and surface of the portion which is common to both.

$$Ans. \tfrac{16}{3}r^3 = \text{vol.}$$
$$8r^2 = \text{surface.}$$

Ex. 212. The axis of a given cylinder passes through the centre of a given sphere; determine the volume of the portion common to both.

Ex. 213. Required the diameter of an auger that, passing through the centre of a sphere, shall cut away *two thirds* of the volume of the sphere.

$$Ans. \text{ Diam. of the auger} = 1.4412R,$$

where R denotes the radius of the sphere.

THE END.

BOOKS FOR SCHOOLS AND COLLEGES

PUBLISHED BY

HARPER & BROTHERS, FRANKLIN SQUARE, N. Y.

☞ HARPER & BROTHERS *will send any of the following Works by Mail to any part of the United States, postage prepaid, on receipt of the Price.*

For a full Descriptive List of Books suitable for Schools and Colleges, see HARPER'S CATALOGUE *and* TRADE-LIST, *which may be obtained gratuitously, on application to the Publishers personally, or by letter, enclosing Six Cents in Stamps.*

Alford's Greek Testament. The Greek Testament : with a Critically Revised Text ; a Digest of various Readings ; Marginal References to Verbal and Idiomatic Usage ; Prolegomena ; and a Critical and Exegetical Commentary. For the Use of Theological Students and Ministers. By HENRY ALFORD, B.D., Dean of Canterbury. Vol. I., containing the Four Gospels, 8vo, Cloth. (*Vol. II. in preparation.*)

Alford's English Testament with Notes. The New Testament for English Readers ; containing the Authorized Version, with Marginal Corrections of Readings and Renderings, Marginal References, and a Critical and Explanatory Commentary. By HENRY ALFORD, B.D., Dean of Canterbury. 8vo. (*In Press.*)

Andrews's Latin English Lexicon, founded on the larger German-Latin Lexicon of Dr. WM. FREUND. With Additions and Corrections from the Lexicons of Gesner, Facciolati, Scheller, Georges, &c. Royal 8vo, Sheep extra.

Abercrombie on the Intellectual Powers. With Questions. 18mo, Cloth.

Abercrombie on the Philosophy of the Moral Feelings. With Questions. 18mo, Cloth.

Alison on Taste. Essays on the Nature and Principles of Taste. Edited for Schools. By Prof. MILLS. 12mo, Cloth.

Anthon's Latin Lessons. Latin Grammar, Part I. Containing the most important Parts of the Grammar of the Latin Language, together with appropriate Exercises in the translating and writing of Latin. 12mo, Sheep.

2   *Harper & Brothers' Books for Schools and Colleges.*

Anthon's Latin Prose Composition. Latin Grammar, Part II. An Introduction to Latin Prose Composition, with a complete Course of Exercises, illustrative of all the important Principles of Latin Syntax. 12mo, Sheep.

>A KEY TO LATIN COMPOSITION may be obtained by Teachers. 12mo, Half Sheep.

Anthon's Zumpt's Latin Grammar. From the Ninth Edition of the Original. By LEONARD SCHMITZ, Ph.D. 12mo, Sheep.

Anthon's Zumpt's Latin Grammar, Abridged. 12mo, Sheep.

Anthon's Latin Versification. In a Series of Progressive Exercises, including Specimens of Translation from the English and German Poetry into Latin Verse. 12mo, Sheep.

>A KEY TO LATIN VERSIFICATION may be obtained by Teachers. 12mo, Sheep,

Anthon's Latin Prosody and Metre. 12mo, Sheep.

Anthon's Cæsar. Cæsar's Commentaries on the Gallic War, and the First Book of the Greek Paraphrase; with English Notes, Critical and Explanatory, Plans of Battles, Sieges, &c., and Historical, Geographical, and Archæological Indexes. Maps, Plans, Portrait, &c. 12mo, Sheep.

Anthon's Æneid of Virgil. With English Notes, Critical and Explanatory, a Metrical Clavis, and an Historical, Geographical, and Mythological Index. Portrait and many Illustrations. 12mo, Sheep.

Anthon's Eclogues and Georgics of Virgil. With English Notes, Critical and Explanatory, and a Metrical Index. 12mo, Sheep.

Anthon's Sallust. Sallust's Jugurthine War and Conspiracy of Catiline. With an English Commentary, and Geographical and Historical Indexes. Portrait. 12mo, Sheep.

Anthon's Horace. The Works of Horace. With English Notes, Critical and Explanatory. A new Edition, corrected and enlarged, with Excursions relative to the Vines and Vineyards of the Ancients; a Life of Horace, a Biographical Sketch of Mæcenas, a Metrical Clavis, &c. 12mo, Sheep.

Anthon's Cicero's Select Orations. With English Notes, Critical and Explanatory, and Historical, Geographical, and Legal Indexes. An improved Edition. Portrait. 12mo, Sheep.

Anthon's Cicero's Tusculan Disputations. With English Notes, &c. 12mo, Sheep.

Anthon's Cicero de Senectute, &c. The De Senectute, De Amicitia, Paradoxa, and Somnium Scipionis of Cicero, and the Life of Atticus, by Cornelius Nepos. With English Notes, &c. 12mo, Sheep.

Anthon's Cicero de Officiis. M. T. Ciceronis de Officiis Libri Tres. With Marginal Analysis and an English Commentary. 12mo, Sheep.

Anthon's Tacitus. The Germania and Agricola, and also Selections from the Annals of Tacitus. With English Notes, &c. Revised and Enlarged Edition. 12mo, Sheep.

Anthon's Cornelius Nepos. Cornelii Nepotis Vitæ Imperatorum. With English Notes, &c. 12mo, Sheep.

Anthon's Juvenal. The Satires of Juvenal and Persius. With English Notes, &c., from the best Commentators. Portrait. 12mo, Sheep.

Anthon's First Greek Lessons, containing the most important Parts of the Grammar of the Greek Language, together with appropriate Exercises in the translating and writing of Greek, for the use of Beginners. 12mo, Sheep.

Anthon's Greek Composition. Greek Lessons, Part II. An Introduction to Greek Prose Composition, with a complete Course of Greek Exercises illustrative of all the important Principles of Greek Syntax. 12mo, Sheep.

Anthon's Greek Grammar. For the Use of Schools and Colleges. 12mo, Sheep.

Anthon's New Greek Grammar. From the German of Kühner, Matthiæ, Buttmann, Rost, and Thiersch; to which are appended Remarks on the Pronunciation of the Greek Language, and Chronological Tables explanatory of the same. 12mo, Sheep.

Anthon's Greek Prosody and Metre. For the use of Schools and Colleges; together with the Choral Scanning of the Prometheus Vinctus of Æschylus, and Œdipus Tyrannus of Sophocles; to which are appended Remarks on the Indo-Germanic Analogies. 12mo, Sheep.

Anthon's Jacobs's Greek Reader, principally from the German Work of Frederic Jacobs. With English Notes, a Metrical Index to Homer and Anacreon, and a copious Lexicon. 12mo, Sheep.

Anthon's Xenophon's Anabasis. With English Notes, a Map arranged according to the latest and best Authorities, and a Plan of the Battle of Cunaxa. 12mo, Sheep.

Anthon's Xenophon's Memorabilia of Socrates. With English Notes, the Prolegomena of Kühner, Wiggers's Life of Socrates, &c., &c. Corrected and enlarged. 12mo, Sheep.

Anthon's Homer. The First Six Books of Homer's Iliad, English Notes, a Metrical Index, and Homeric Glossary. Portrait. 12mo, Sheep.

Anthon's Manual of Greek Antiquities. From the best and most recent Sources. Numerous Illustrations. 12mo, Sheep.

Anthon's Manual of Roman Antiquities. From the most recent German Works. With a Description of the City of Rome, &c. Numerous Illustrations. 12mo, Sheep.

Anthon's Manual of Greek Literature. From the earliest authentic Periods to the close of the Byzantine Era. With a Critical History of the Greek Language. 12mo, Sheep.

Anthon's Smith's Dictionary of Antiquities. A Dictionary of Greek and Roman Antiquities, from the best Authorities, and embodying all the recent Discoveries of the most eminent German Philologists and Jurists. American Edition, corrected and enlarged, and containing also numerous Articles relative to the Botany, Mineralogy, and Zoology of the Ancients. Royal 8vo, Sheep.

Smith's Antiquities, Abridged by the Authors. 12mo, Half Sheep.

Anthon's Classical Dictionary. Containing an Account of the principal Proper Names mentioned in Ancient Authors, and intended to elucidate all the important Points connected with the Geography, History, Biography, Mythology, and Fine Arts of the Greeks and Romans, together with an Account of the Coins, Weights, and Measures of the Ancients, with Tabular Values of the same. Royal 8vo, Sheep.

Anthon's Smith's New Classical Dictionary of Greek and Roman Biography, Mythology, and Geography. Numerous Corrections and Additions. Royal 8vo, Sheep.

Anthon's Latin-English and English-Latin Dictionary. For the use of Schools. Chiefly from the Lexicons of Freund, Georges, and Kaltschmidt. Small 4to, Sheep.

Anthon's Riddle and Arnold's English-Latin Lexicon, founded on the German-Latin Dictionary of Dr. C. E. GEORGES. With a copious Dictionary of Proper Names from the best Sources. Royal 8vo, Sheep.

Anthon's Ancient and Mædieval Geography. For the use of Schools and Colleges. 8vo, Cloth.

Barton's Grammar. With a brief Exposition of the chief Idiomatic Peculiarities of the English Language. With Questions. 16mo, Cloth.

Beecher's (Miss) Physiology and Calisthenics. Over 100 Engravings. 16mo, Cloth.

Boyd's Eolectic Moral Philosophy; for Literary Institutions and General Use. 12mo, Cloth.

Boyd's Rhetoric. Elements of Rhetoric and Literary Criticism, with copious Practical Exercises and Examples: including, also, a succinct History of the English Language, and of British and American Literature, from the Earliest to the Present Times. On the Basis of the recent Works of ALEXANDER REID and R. CONNELL; with large Additions from other Sources. 12mo, Half Roan.

Burke on the Sublime and Beautiful. 12mo, Cloth.

Butler's Analogy. By EMORY and CROOKS. Bishop Butler's Analogy of Religion, Natural and Revealed, to the Constitution and Course of Nature. With an Analysis by the late ROBERT EMORY, D.D., President of Dickinson College. Edited, with a Life of Bishop Butler, Notes and Index, by Rev. G. R. CROOKS, D.D. 12mo, Cloth.

Butler's Analogy. By HOBART. Prepared for Schools, by CHARLES E. WEST. 18mo, Cloth.

Butler's Analogy. Edited by HALIFAX. 18mo, Cloth.

Buttmann's Greek Grammar. For the use of High Schools and Universities. Revised and Enlarged. Translated by EDWARD ROBINSON, D.D., LL.D. 8vo, Sheep.

Calkins's Object Lessons. Primary Object Lessons for a graduated Course of Development. A Manual for Teachers and Parents, with Lessons for the Proper Training of the Faculties of Children. By N. A. CALKINS. Illustrations. 12mo, Cloth.

Campbell's Philosophy of Rhetoric. 12mo, Cloth,

Clark's Elements of Algebra. 8vo, Sheep.

Collord's Latin Accidence. Latin Accidence and Primary Lesson Book; containing a full Exhibition of the Form of Words, and First Lessons in Reading. By GEORGE W. COLLORD, A.M., Professor of Latin and Greek in the Brooklyn Collegiate and Polytechnic Institute. 12mo, Sheep.

Combe's Physiology. With Questions. Engravings. 18mo, Cloth.

Comte's Philosophy of Mathematics. Translated from the Cours de Philosophie Positive, by Prof. W. M. GILLESPIE, A.M. 8vo, Cloth.

Crabb's English Synonyms. English Synonyms Explained. With copious Illustrations and Explanations, drawn from the best Writers. By GEORGE CRABB, M.A. 8vo, Sheep.

Curtius and Smith's Series of Greek and Latin Elementary Works. Revised and Edited by Professor HENRY DRISLER, of Columbia College, N. Y.:

    1. **Principia Latina**, Part I. A first Latin Course, by Dr. W. SMITH, LL.D., &c. 12mo, Cloth.

    2. **Principia Latina**, Part II. Latin Prose Reading-Book, by Dr. W. SMITH (containing Viri Romæ, &c.). 12mo, Cloth.

    3. **Principia Latina**, Part III. (IV.) Latin Prose Composition, Rules of Syntax, with copious Examples, Synonyms, &c., by Dr. W. SMITH.

www.ingramcontent.com/pod-product-compliance
Lightning Source LLC
Chambersburg PA
CBHW022042230426
43672CB00008B/1046